THE PREHISTORY
OF
UGANDA PROTECTORATE

THE PREHISTORY
OF
UGANDA PROTECTORATE

by

T. P. O'BRIEN, F.G.S.

With a Chapter on
THE PLEISTOCENE SUCCESSION
by
J. D. SOLOMON, B.A., Ph.D., F.G.S.

and an Appendix on
THE MAMMALIAN FOSSILS
by
A. TINDELL HOPWOOD, D.Sc., F.L.S.
Department of Geology, British Museum (Natural History)

CAMBRIDGE
AT THE UNIVERSITY PRESS
1939

CAMBRIDGE
UNIVERSITY PRESS

32 Avenue of the Americas, New York NY 10013-2473, USA

Cambridge University Press is part of the University of Cambridge.

It furthers the University's mission by disseminating knowledge in the pursuit of education, learning and research at the highest international levels of excellence.

www.cambridge.org
Information on this title: www.cambridge.org/9781107419155

© Cambridge University Press 1939

First published 1939
First paperback edition 2014

A catalogue record for this publication is available from the British Library

ISBN 978-1-107-41915-5 Paperback

CONTENTS

*available for download from www.cambridge.org/9781107419155

v

ILLUSTRATIONS

PLATES

vii

Illustrations

FIGURES

Illustrations

Illustrations

PREFACE

THE geological and archaeological results described in this work were obtained during a season of some eighteen months in Uganda Protectorate. A previous visit to the country and a meeting with Mr E. J. Wayland, Director of the Geological Survey, had suggested that Uganda was a most promising region in which to extend the investigations on East African archaeology, already carried so far in Kenya Colony. As Uganda lies just west of Kenya and yet possesses such marked differences in climate and topography, it seemed to offer the best chance of checking and supplementing the Kenya sequence, particularly in regard to the regional significance of the Kenya cultures. As will be shown, Uganda proved that the latter must not be regarded any longer as ubiquitous in East Africa, but that they were often only local cultures.

I cannot leave without record the disinterested way in which Mr Wayland welcomed the expedition to the country in which he had priority of research, and his generous encouragement. Similarly, he most kindly placed at our disposal the results of his own investigations which were invaluable in providing us at the start with information as to possible sites and lines of enquiry. I gratefully acknowledge also his kindness in lending us tools and for providing us with accommodation for ourselves, equipment and specimens.

But above all must I acknowledge Mr Wayland's scientific spirit in allowing me to publish our results both in advance of, and separately from, his own, although it was assumed originally that both would appear together.

I am also most grateful to many people, too numerous to mention individually, who gave us hospitality, information and other help. My thanks are also due to the Kenya and Uganda Railways and Harbours for allowing our specimens to be sent to the coast at special rates.

To the authorities of the Musée du Trocadero, Paris, I owe a special debt for their kindness in giving us space in which to work out our material, at a time when they themselves were short of room, and also to M. L'Abbé Breuil, Mr and Mrs Harper Kelley and Mrs Bowler-Kelley.

Preface

I wish especially to thank Mr M. C. Burkitt for his unfailing encouragement and advice on many problems.

To Dr A. T. Hopwood I offer my grateful thanks for undertaking the identification of the fossil remains and for writing Appendix B for this book.

I am particularly indebted to Dr J. D. Solomon, not only for his invaluable work on the geological sequence and for writing Chapter III, but also for much that he has taught me, either by example, or in the course of many discussions about geological matters. In fairness to him I must add that he must not be held responsible for some of my statements about things that he was unable to verify for himself, or for others made when I felt obliged to amplify his ideas in connection with the archaeological side of our studies.

I offer my grateful thanks to Mr H. J. H. Drummond for his most generous financial support without which our season in Uganda would have been halved. I also acknowledge with thanks many other donations to the cost of the expedition from friends interested in the work.

I also wish to thank Mr C. O. Waterhouse of the British Museum for his extremely efficient drawings of the majority of the stone tools figured here.

Finally, but above all, do I thank my wife for her willing duties in camp maintenance, marking and packing of specimens and other monotonous work, for her help in writing this book and for being at all times a cheerful and efficient colleague.

T. P. O'B.

1939

CHAPTER I

The Country

UGANDA is a small country, nearly one-seventh of whose surface is occupied by water, leaving a land area rather less than that of England and Scotland together.

It lies on the Equator and is bounded on the north by the Anglo-Egyptian Sudan, on the south by the Belgian mandate of Ruanda-Urundi and by Lake Victoria, with Tanganyika Territory beyond, on the east by Kenya Colony and on the west by the Belgian Congo.

The central part of the country is a plateau, dissected by many rivers, swamps and lakes, of a general level of 4000 ft. above the sea. East, west and south-west are high mountains, while to the north, beyond Lake Kioga, the land slopes gently down, with scattered hills standing out of the plain, to the Sudan, at a little over 2000 ft. above sea-level.

The mountains and part of the central plateau round the lakes are forested and very wet. Most of the central plateau is savannah country, with a good deal of cultivation, while the south-western uplands are rolling, grassy hills except where they rise to mountains of forest height. The last two areas receive a medium rainfall. The Western Rift Valley, with Lakes George, Edward and Albert, is rather dry, with scrubby bush, while Karamoja, the north-eastern corner of the Protectorate, is very arid indeed, owing to the strongly seasonal nature of the rainfall, which is often violent, but soon wasted by rapid evaporation and surface run-off.

The present topography of Uganda is very largely of recent geological age. It is a three-peneplain[1] topography, of which the second peneplain is widely represented by flat-topped residuals between valleys whose existence is due to uplift movements of post-Miocene age. These movements seem to have been part of the general uplift of eastern Africa, whose more striking results were the rift valleys and associated flexures.

[1] But see Solomon, p. 17, who regards Peneplain I as the westerly prolongation of Peneplain II, or the Buganda Peneplain.

Apart from faulting, the most important effect of these movements was rejuvenation; valleys were carved out and some of them had reached an advanced stage of maturity by Upper Pliocene times.[1]

Several of the major rivers of Uganda occupy deep, mature valleys of the "two-way" type; that is, their headwaters rise in shallow swamp-divides, from which the almost imperceptible flow is in two directions, one towards the Albertine Rift[2] and the other towards either Lake Victoria or Lake Kioga. The Kafu, in the central region, and the Kagera, in the extreme south, are examples. In all cases, these swamp-divides are situated along a south-west, north-east axis, between twenty and thirty miles from the Western Rift and parallel to it, and they coincide with the general position of an uplifted area bordering it. It is clear that this relatively local uplift is more recent than the main, regional uplift of the Miocene peneplain, and is intimately connected with the genesis of the Western Rift Valley, which itself was the result of stresses imposed during the regional uplift.

Wayland was the first to recognise compression as the cause of the Albert Rift, in particular, as opposed to the general tensional theory put forward by Gregory. Bailey Willis[3] also examined the area during a recent visit to Uganda, and is in agreement with Wayland on this point. Proof of the compression is seen in overthrusts, first postulated by Wayland and actually recognised by Bailey Willis, and, in a general way, in the undoubted uplift on either side. Speaking of this compressive action in relation to the geology of the valley, Bailey Willis says:

The escarpment is there (Kibero, where Willis observed the overthrust) *about a thousand feet high . . . and is the central section of the main, eroded fault scarp. It is . . . modified by benches that presumably represent successive steps and*

[1] These mature valleys would appear to have been the narrow equivalents of the third Peneplain which is represented by the great plains of Busoga and, in a lesser degree, in Bunyoro.

[2] The term "Western Rift Valley", including Lakes Albert, Edward and George, is rather a loose one, for the Valley is not entirely rifted, but contains areas that have been downwarped and never subsequently faulted. Lakes Edward and George, for instance, occupy a depression which shows only little evidence of faulting on its eastern side, though the Congo side appears to be bounded by a definite fault-scarp. For the sake of convenience, however, both these areas, the Edward-George and the Albert valleys, are here referred to as the Albert or Western Rift Valley.

[3] Bailey Willis, *East African Plateaus and Rift Valleys*, 1936, pp. 47–8.

may be upthrusts or secondary, gravity step faults. The exposure of the ramp or overthrust is an accident of erosion, where a ravine cuts across the scarp and the section reveals the rocks....Their ancient structures were imposed under excessive confining pressure and are characterised by recrystallisation and flowage, rather than by fracture. The thrust on the other hand, is a curved shear, which has resulted in much fracturing and displacement of blocks, as under light load. It could have been produced only in rocks near the surface, that is only when erosion had removed the burden of many thousand feet. Occurring, as it does, in a position where it ramps toward the free face of the steep escarpment it is mechanically connected with that freedom from resistance and presents the action of horizontal pressure from the plateau towards the trough. The ramp thus serves to confirm the inference logically drawn by Wayland from the general relations of the Albert trough to the updoming of Ruwenzori.

There is strong reason for believing that the earliest compressive stresses, which developed along what is now the Albert Rift Valley, resulted first in a simple downwarp or trough. Remnants of the down-tilted peneplain can be seen at a number of places, particularly along the south-east edge. It is clear that the first actual fault of rift valley magnitude truncated the sides of what Wayland has called the "hanging topography" of the down-tilted Miocene peneplain.

While the chief result of the lateral compression was the formation of a long trough, which subsequently developed into the Albert Rift Valley, some of the stresses were also from below, and caused the uplift of a long tract of country bordering the depressed trough. To this uplift may be ascribed the first reversal of the long, westerly flowing rivers. Those parts of their valleys which lay on the trough side of the uplifted zone were either rejuvenated or else became insignificant streams of little erosive power, according as to whether their headwaters drained a wide area within the uplifted zone or not. An instance of the latter is the Nkussi, and of the former, the Muzizi. This river, with a comparatively short course, appears always to have drained an enormous area north-east of Ruwenzori, and its whole catchment area was rejuvenated at the time of the uplift movements. Erosion seems to have kept pace with uplift, so that the valley was always more or less graded to the floor of the trough, for, to-day, remnants of the earliest lacustrine sediments, the Kisegi Beds, are to be seen in the floor of the valley, more than 800 ft. below the level of the peneplain, but above the falls by which the river now enters the Rift Valley, and which were caused by the major faulting of post-Kisegi times. Kisegi Beds are also to be seen down in the Albert Rift, more than a thousand feet below the Muzizi valley.

In the case of another river, the Kagera, it is also true that the western end, which to-day flows into Lake Edward (at the southern extremity of the Western Rift), runs through very high country, mostly old remnants of Peneplain I, and possesses a very large catchment area, like the Muzizi. Consequently, when the uplift of the edge of the Albert trough took place, that part of the uplifted area containing the catchment zone was tremendously rejuvenated westwards towards the depression, though there may have been, and very likely was, reversal on the eastern side. To-day, this western end of the river enters the Edward depression through a very deep and obviously ancient valley which has never been truncated by rifting, as in the case of the Muzizi, because faulting has not played so great a part in the formation of the Edward depression as in the Albert basin, particularly along its eastern side.

Lake Victoria and Lake Kioga both owe their origin to this secondary uplift along the edge of the Albertine trough, first, because the movement caused the creation of shallow basins, and, secondly, because these depressions were filled by the waters of such rivers as were reversed by the movement.

The events so far described took place in pre-Pleistocene days, probably during the Pliocene, and have little connection with the Pleistocene succession except in so far as they modified the Miocene levels and created new base-levels, thus directing the course of the early Pleistocene rivers.

There is, however, much evidence of later movements also, which affected these rivers, and one such movement, at least, appears to have been a tilt from north-east to south-west which reversed the easterly flowing rivers back to the Albertine Rift in human times. This may have been due to faulting within the Rift, though this seems unlikely, as the main fault, which produced the Albert Escarpment as we know it to-day, is demonstrably older than any of the Pleistocene deposits in the

4

Rift. It may be that this tilt was the result of slipping at depth along the plane of the older faults, or overthrusts.

Wayland ascribes this river reversal to the rise of the lakes and their flooding up the rivers due to pluvial conditions, and not to tilting, but, apart from lack of evidence of pluviation at this period, there is the evidence of tilted beaches in support of a north-east to south-west movement at this time.

The final tilt was back towards the north-east, reversing the direction of flow once more and leaving the rivers and valleys much as we know them to-day.

CHAPTER II

Research in Uganda prior to 1935

In this chapter, I shall do no more than summarise, as briefly and clearly as possible, the results achieved by Mr E. J. Wayland in the course of his 15 years' work up to the end of 1934. I shall present only his own views and data, as published from time to time, leaving all discussion until the appropriate chapter. Solomon will, of course, reconsider the evidence of pluviation in East Africa when dealing with his own results in Chapter III.

PLEISTOCENE CLIMATES IN UGANDA

Wayland's attitude and approach to the problems of Pleistocene climates in East Africa is best summed up in his own words:[1]

The very wide distribution of perched gravels in Uganda, and their not unusual occurrence at exceptionally high levels above present day streams or valley bottoms, attracts the attention of a geologist at once; so too does the ubiquity of stone age artifacts...and often enough it is in gravels that these have been preserved.... Similar...occurrences and associations in Ceylon were, before the war, partly responsible for the writer adopting Brooks' view with regard to the major climatic events of the Pleistocene. Brooks points out that large ice sheets promote permanent anticyclones above them, and that precipitation within such anticyclonic areas is slight, the deficit being made up in non-glaciated regions. Applying this to the conditions of the Glacial Period, Brooks contends that the existence of vast ice sheets in the higher latitudes and heavy rainfall elsewhere is at once in accordance with the demands of meteorological theory and the facts of geological evidence. Accepting this, the present writer saw in the widespread Pleistocene gravels of Ceylon additional support of Brooks' view, and it required no effort of imagination similarly to account in part at least, for many of the Uganda gravels.

As Wayland goes on to record, he came to the conclusion, in 1919, that Lake Victoria had been higher in level, "*probably in consequence of the meteorological conditions of the Ice Age*". By 1922, he had evidence of

---

[1] *Summary of Progress of the Geol. Survey of Uganda*, 1919–29, pp. 37 *et seqq.*

three pluvial periods, "*but could discover no evidence of a fourth which Brooks' hypothesis, combined with the generally accepted fourfold division of the Ice Age...seemed to demand*". Later, the evidence seemed to be partly interpretable in another way[1] but, by 1926, Wayland "*was still strongly inclined towards a three-glacio-pluvial hypothesis*".

Leakey's and Nilsson's independent results in 1927 appeared to confirm Wayland's threefold pluvial scheme, but, later in the same year, further work in Uganda caused him to effect a rearrangement, as he no longer felt that the Uganda pluvials each coincided with one main ice advance (Günz-Mindel, Riss and Würm) but, if really contemporary ice advances, "*then rather with Günz-Mindel, Riss-Würm and a post-glacial event, possibly the Buhl stadium*".[2]

Until 1929, Leakey still followed Wayland's older correlation between the three main European glaciations and the three East African pluvials (*sic*), but, in that year, as the result of Solomon's collaboration, the third "pluvial" was considered, as really a post-pluvial wet phase, possibly corresponding, as Wayland thought, with Buhl.

In 1929 also, Nilsson published his preliminary paper on his own detailed work on the Kenya lake beds and East African mountains, where he obtained, apparently, the most convincing evidence of three major pluvials of glacial age and two more of post-glacial age. However, Wayland's, Leakey's and Solomon's agreement on two major pluvials, each corresponding to two European glacials, tended to overshadow Nilsson's results and received further support from Brooks' independent opinion that

the most we can do is to separate the glacial period into three subdivisions, a first glacial period corresponding with the Günz and Mindel glaciations in the Alps, a long "Interglacial period" which was the Mindel-Riss, and a second glacial period including the Riss and Würm stages.[3]

Still further support for the Wayland-Leakey-Solomon glacio-pluvial theory was forthcoming when Dr (now Sir George) Simpson, Director of the Meteorological Office, London, gave his lecture on Past Climates.[4] In this, Simpson demonstrated that

[1] *Man*, 124, 1924.   [2] *Sum. Prog. Geol. Surv. Ug.* 1919–29, p. 38.
[3] Brooks, *Climate through the Ages*.   [4] *Alexander Pedler Lecture*, British Science Guild.

from the point of view of past climates two meteorological elements are outstanding, temperature and precipitation.

Stated broadly, two cycles of increased solar radiation and, thereby, of temperature, over the whole world, must have led to increased evaporation and, consequently, precipitation. This, necessarily, led to the feeding and growth of existing ice-sheets round the poles and to pluvials in equatorial and subtropical zones. Continued increase of solar radiation in each rising cycle led to excessive temperatures and, eventually, to the melting of the ice-sheets and to the existing glacial climate giving place to a warm, wet interglacial. In the tropical zones, however, precipitation would be continuous. The second glaciation of each radiation cycle would gradually come on during the decline of the radiation cycle, only to give place to a cold, dry interglacial stage when precipitation slackened off all over the world. This interglacial would thus coincide with the interpluvial in unglaciated regions.

Simpson's theory seemed to remove the difficulty, fully realised by Wayland, that there only seemed to be evidence of two main pluvials in East Africa, to equate with four European glacials, and Simpson's main conclusion, that

each pair of glacial periods, with the intervening warm wet interglacial period coincides with a pluvial period in unglaciated regions,

from henceforth received Wayland's unqualified support. From 1929 onwards, with one exception, there was no significant change in Wayland's views on this subject. This single exception was the emphasis laid, in 1932, on the existence of *intrapluvial* oscillations within the pluvials. Until that date, there is, so far as I know, no mention of these phenomena in Wayland's published views, but, in 1932, their existence, both in Uganda and at Oldoway, is almost categorically stated. Dealing first with Uganda, Wayland writes:[1]

It was discovered in Uganda, and subsequently in Kenya, that Pluvial I had two rainfall peaks. It is now found that in the Kagera valley there is evidence of a similar climatic oscillation in the latter part of Pluvial II.

[1] *Annual Report of the Geological Survey of Uganda*, 1932, p. 16.

Then, speaking of Oldoway, which he had just visited, he writes:[1]

I contend that a stratigraphical break exists between Beds III and IV, that this is associated with a climatic oscillation, and that it is the only stratigraphical break of any consequence. . . .

At that time he felt the break to be that between Pluvials I and II, but, later,[2] he stated that

my own work in Uganda shortly revealed this same climatic episode as an oscillation within a pluvial; not in Pluvial I, however, but in Pluvial II.

As far as one can judge, the evidence in support of the intrapluvial in Pluvial I was provided by local, though not intense, soil reddening, selenite beds within Pluvial I deposits and by talus accumulations of this date, separating water-laid deposits.

The evidence for the intrapluvial in Pluvial II was much more marked, according to Wayland, as it led to much greater soil reddening, etc., though it was in no sense as marked as in the great Interpluvial.

The intensity of the latter was such, in Wayland's opinion, that it caused the almost complete drying up of Lake Albert and the extinction of some animal forms by the time that Pluvial II began. It was responsible for the important Kaiso Bone Beds, remains from which were described, in 1926, by British Museum authorities.[3]

It is evident that, by 1933, probably as a result of his visit to Oldoway, Wayland realised that it was impossible to effect a close correlation between his two major pluvials and Leakey's Kamasian and Gamblian periods, and he had already shown[4] how there seemed to be some disparity between the cultures, claimed to be of the same periods, on each side of Lake Victoria. In his opinion, Leakey and his colleagues had placed the Interpluvial too far up in the scale, and much that they included in their first Pluvial actually belonged to the second. The stratigraphical break that Wayland observed at Oldoway (marked by the reddened Bed III) was, at first, thought by him to mark the Interpluvial, but, as already recorded above, he very soon found that it was,

---

[1] *Loc. cit.* p. 14.

[2] *Rifts, Rivers, Rains and Early Man in Uganda*, p. 343. *J.R.A.I.* vol. LXIV, July–December, 1934.

[3] *Occ. Paper, No.* 2, Geol. Surv. Uganda, 1926.

[4] *Antiquity*, 1932.

apparently, the same as that displayed in the Kagera valley, Uganda, where it was clear that it occurred in Pluvial II. As he remarks,[1]

it follows that much of the Kamasian belongs, not as the East African Archaeo-logical Expedition claims, to the first Pluvial, but to the second.

In point of fact, as perusal of Wayland's 1934 paper shows, the *whole* of Leakey's Kamasian, as displayed at Oldoway, beginning with the pre-Chellean implementiferous Bed I, belongs to Wayland's Pluvial II, though it is obvious that some of the earlier Kenya deposits, not implementiferous and mainly volcanic tuffs, occurring round the Rift Valley, were most probably formed at the same period as Wayland's Pluvial I. At the same time, the implications of Simpson's theory of essential contemporaneity between the pairs of glacials and the major pluvials, led to still greater discrepancies between the Kenya and Uganda dating. While Leakey, on the one hand, made his two pluvials —Kamasian and Gamblian—correspond (probably) with Günz-and-Mindel and Riss-and-Würm respectively, thereby bringing the archaeological sequence more into line with the European dating, Wayland's interpretation led him to correlate as follows:[2]

assuming that the African pluvials are to be equated with the glacials of higher latitudes, and having recourse for the moment to Alpine nomenclature, the Chelleo-Acheulean culture of Kenya and Uganda belongs not to the close of Mindel times, but to the end of the Riss and the beginning of Würm days,

an assumption widely at variance with the known facts in Europe.

With regard to the intensity of the pluvials, Wayland has stated[3] quite clearly that, in his opinion, there was nothing "*catastrophic in the climatic phases of the Pleistocene in Central Africa*", nor did he see reason

to suppose that precipitation, even during the first pluvial, was excessive for the tropics as we know them in some parts today; nor do I think that desiccation was sustained or intense during the intrapluvials.

ARCHAEOLOGY

Wayland's own account of his archaeological investigations, which was published in 1934,[4] makes it unnecessary to do more here than indicate

---

[1] *R.R.R.E.M.U.* p. 343.   [2] *Loc. cit.*
[3] *Loc. cit.* p. 348.   [4] *Op. cit.*

briefly the results that had been achieved up to the time of our arrival, in September 1934. Ever since 1919, when he was appointed Director of the Geological Survey of Uganda, Wayland had made full use of his opportunities for archaeological research and had collected assiduously at the same time as he investigated the Pleistocene history of the country.

The *Annual Reports* of the Geological Survey form the chief source of published information on the Stone Age cultures, but numerous other papers in various journals also attest Wayland's energy in this field, and the greatest credit is due to him for his splendid work, much of which has received too little attention.

In 1919, the two cultures which, to-day, are still, in part, the earliest known in East Africa, were discovered in the gravels of the Kafu and Muzizi rivers respectively. They were the Kafuan and Oldowan, both of which names were applied later, though the pre-Chellean age of the cultures was recognised by Wayland at the time.

In the same year, Chelleo-Acheulean, Levalloisian, or "Mousterian" and microlithic industries were found at various places.

In 1920, further finds of microlithic industries were made at other sites, and Mr W. C. Simmons, also of the Geological Survey, discovered an industry which we now know to be the best representative of the Kenya type of Still Bay in Uganda.

The Sango Hills, at the mouth of the Kagera River, provided a series of large stone tools which, at a later date, Wayland was to regard as a form of Mousterian and, still later, as a specialised form of Chelleo-Acheulean, leading to the Mousterian (cf. Levalloisian).

In 1924, Chelleo-Acheulean tools were found in lateritised lake terraces on Buvuma Island, adding further data to the accumulating evidence that Man had witnessed many important physiographical changes in Uganda.

In 1925, large collections were made in Karamoja, in the north-east of the Protectorate. These included Mousterian (cf. Levalloisian), Aurignacian (cf. East African Capsian, or derivative industries) and developed Kafuan.

In 1926, a silted-up granite water-hole at Magosi, in Karamoja, was trenched and pitted, providing an interesting industry of Early Wilton facies, containing Still Bay elements.

In 1930, the extremely rich neighbourhood of Nsongezi, in Ankole District, was visited, and collections of Chelleo-Acheulean tools were made.

In 1932, work in the Nsongezi and Kagera valley area was carried on, and further collections were made at many other sites in the following years.

THE CULTURES

Wayland's conclusions on the cultures up to the end of 1934 are given here, in rather more detail, in order to show the position at that time and so that the changes and additions consequent upon our own work may be fully understood.

Until Leakey discovered the pre-Chellean pebble-culture known as Oldowan, at Oldoway, Tanganyika, the pre-Chellean age of the two Uganda pebble-cultures was based only on their very primitive appearance. Unfortunately, neither of them had been found in deposits which yielded a sequence of Chelleo-Acheulean types, upon which to base any comparisons of age or relationships; nor can it be said that the geology of the terrace sites, admirably as it was studied, provided conclusive evidence of the age of the contained tools. This is evident from the fact that, in the *Annual Report* for 1927, after he had paid a visit to Leakey's sites in Kenya and after two seasons' intensive research into the history of the Kafu River, Wayland regarded the *pre-Chellean* Kafuan as probably the direct ancestor of the Sangoan culture—itself considered as contemporary with Leakey's Kenya Aurignacian from such sites as Gamble's Cave.

As regards the supposedly younger series, at the Muzizi River, perusal of all the available published data gives no further suggestion of its age than the early implication of being also pre-Chellean.

By 1932, after Oldoway had been visited, the Kafuan is placed right at the base of the culture sequence, in the early part of Wayland's Pluvial I, followed by other pre-Chellean industries and then by true Chellean, which was thought to have died out by the time of the Kaiso Interpluvial, which was, in turn, followed by Acheulean in the early part of Pluvial II.

By 1934, however, the Kafuan alone was thought to belong to

Pluvial I and the early part of Pluvial II, while the hand-axe culture came entirely in Pluvial II.

With all respect to Wayland, I must record my opinion that the ages of the early cultures of Uganda were still largely a matter for speculation at the end of 1934, as was the correlation of the Kafu River terraces with the Pluvial I.

Tools of Chelleo-Acheulean facies had been found in Uganda ever since 1919, and, in that year, Wayland had claimed that this specific, if comprehensive, complex was present in the country; yet it was not until some years had passed that he was prepared to admit the existence of Chellean and Acheulean cultures as distinct from the Sangoan which, at least until 1927, was regarded as a curious specialisation of the local Mousterian, containing large *coups de poing* and picks.[1]

By 1929, however, Wayland had recognised the very mixed nature of the Sangoan and records the probability that pre-Chellean and Chelleo-Acheulean industries were represented in the Sangoan portmanteau. By 1930, after preliminary work at Nsongezi, there was no further doubt, and Wayland remarks[2] that

it was becoming apparent that whatever the typological position of the Sangoan might be, my original claim, made in 1919, *that a true Chelleo-Acheulean facies occurred must, after all, be sustained.*

By the end of 1932, after his visit to Oldoway, Wayland made a correlation between his M-Horizon Intrapluvial (in Pluvial II) and its culture, and a climatic oscillation recorded in the Oldoway series which he regarded as similar—the Red Bed III, which contained an industry then considered to be transitional between Chellean and Acheulean. According to Leakey now, this stage of culture is Early Acheulean.

Reference to the Table of Cultures published in 1934[3] will show that, by this time, Wayland believed the Chelleo-Acheulean to be distinct from, though contemporary with, the Sangoan. Thus, he appears to have dropped the idea that the Sangoan was a mixture of Chellean, Acheulean and Mousterian. Also, while the Chelleo-Acheulean com-

[1] *A.R.G.S.U.* 1927, p. 34.
[2] *R.R.R.E.M.U.* p. 342.
[3] *Op. cit.*

plex was thought to be a culture of the valleys, the Sangoan was regarded as a hill-culture. Wayland says:[1]

The Sangoan facies out-ran, so to say, the Acheulean proper and replaced it, only to develop, or so it would seem from what one knows at present, into the Mousterian by the dropping out of core tools (which also diminished in size) pari passu with the development of the flake technique.

Thus, in his table, the Sangoan is shown to evolve into the Mousterian and, thence, into the Still Bay.

Of the so-called Aurignacian, reported from the extreme northeastern area, bounding Kenya, Wayland admitted that it was a foreign influence, "*decidedly rare elsewhere in Uganda*".[2] Judging, presumably, by the Kenya data, the Aurignacian was considered as having given rise to microlithic industries like the Magosian and Wilton.

This concludes the brief sketch of Wayland's major archaeological results and his main conclusions. Table I (after p. 316), from his paper of 1934, shows the extent to which we are indebted to him for his laborious and painstaking pioneer work. Though we differ, sometimes considerably, on many points, that is immaterial and no less credit is due to him on that account.

[1] *R.R.R.E.M.U.* p. 351.
[2] *Loc. cit.*

CHAPTER III

The Pleistocene Succession in Uganda[1]

THE evidence available for deciphering the sequence of events in Uganda during the Pleistocene period is very scrappy; there is no one locality which provides a clear record of more than a very limited part of the succession of events, and in attempting a synthesis of the geological record it is therefore necessary to take into account the widest possible considerations, particularly in the field of geomorphology. It is also necessary to make use of such correlations between Palaeolithic implements and fauna as have been established in the neighbouring territories, especially at Oldoway and in the Kenya Rift Valley, for in Uganda there is no locality where the two are found together in the same deposits. It follows from these considerations that the work of synthesis is largely interpretational; and, indeed, the present writer does not claim to have discovered the major part of the evidence which he has used in his reconstruction of the sequence of events.

One such reconstruction is already in the field, namely, that of Mr E. J. Wayland, who has been in the country since 1919 and whose diligent and enthusiastic collection of the sparse and widely scattered data available deserves the tribute of all who follow after him.

It is proposed to give here only a brief summary of his ideas, for a fuller account is easily available in his communication of 1934 to the *Journal of the Royal Anthropological Institute*, vol. LXIV, pp. 33 *et seqq.*, and O'Brien also, in the last chapter, gave a historical summary of Wayland's views from 1919 onwards.

Briefly, he considers that two major pluvial periods, each divided into two wet spells separated by an intrapluvial drier episode, occurred during the Pleistocene. He considers that the Albertine Rift was greatly intensified during the Pleistocene epoch, and that the rise and fall of the Uganda lakes, as well as the reversal of many of the rivers, are to be correlated with the climatic sequence.

He refers the older Kafu terraces and the Kisegi Beds in the Albert

[1] By J. D. Solomon, B.A., Ph.D., F.G.S.

Rift to Pluvial I, part 1; the Kaiso Beds and the 50 ft. and "flats" Kafu terraces to part 2 of the same pluvial period, while the Epi-Kaiso Beds and the alluvial deposits of the Kagera valley are attributed to Pluvial II, the M-Horizon rubble of Nsongezi representing the Intrapluvial. Wayland regards the Kaiso Bone Beds of the Albert Rift as marking the Interpluvial between Pluvials I and II.

The present writer, while fully in sympathy with the idea that a well-established climatic sequence in tropical Africa might provide a valuable link with the European ice-ages, has, on examination of the evidence in the field, felt obliged to discard the pluvial hypothesis more or less completely, and has, in consequence, attempted a reconstruction on quite different lines, interpreting the field evidence in terms of regional earth movements rather than of changes of climate. It is not the writer's intention to disparage the very ingenious scheme devised by Wayland, but he has long felt that an alternative one was possible and has here endeavoured to provide it, in order that future workers may be able to select either one in which to fit any fresh evidence that may be found.

The evidence at present available falls into two classes:

    (i) Geomorphological,

    (ii) Stratigraphical.

These are treated separately below; they naturally overlap at times, and such points of contact are dealt with in the text. A general synthesis is attempted in conclusion.

GEOMORPHOLOGICAL EVIDENCE

(a) *The Peneplains*

Wayland has rightly pointed out that the greater part of Uganda is notable for the examples of ancient peneplanation preserved there, which he terms (in chronological order) the Ankole-Kigezi Peneplain, the Buganda Peneplain and a third of Pliocene age. He is disposed to recognise substages of the second erosional phase.

The present writer was unable to see any very clear evidence of more than two principal stages, and suggests tentatively that the Ankole-Kigezi Peneplain may be the westerly prolongation of the Buganda one rather than an older erosion level, for these two do not appear to occur in juxtaposition; their contact might be expected in the country to the west and south-west of Masaka, but, unfortunately, later erosion seems to have removed the pre-Pliocene peneplain levels in that area.

In view of the great perfection of the Buganda peneplanation in the type area of Buganda itself, where it has, as far as I am aware, left no monadnocks, it seems unnatural that the only remnants of it in Ankole should be a few flattened spurs which Wayland assigns to it in that region.

The levels of the Ankole-Kigezi and Buganda peneplains are also quite consistent with the hypothesis that the one is the prolongation of the other; the Buganda hilltop levels rise from between 4000 and 4400 ft. O.D. near Kampala to 4700 near Masaka, and this gradient would take the erosion plane quite naturally into the hilltops near Rakai, which are slightly over 5000 ft. In any case, some discrepancy might be expected owing to warping, and this has certainly affected the Buganda peneplain, as can be especially well seen near Mpigi, where the hilltops are beautiful specimens of the inclined plane.

Thus, in the writer's opinion, the topography of Uganda is chiefly determined by two peneplains; the older was a very perfect one, while the younger never attained anything like completion before the base level of erosion once more suffered alteration. The older peneplain comprises the Buganda and Ankole-Kigezi levels of Wayland, while the younger is the widespread erosion level so obvious in Bunyoro, the northern part of Masaka district and in much of Toro, and which extends far eastwards into Busoga and the Kavirondo Province of Kenya Colony. Indeed, it is very probably the same as the great plain east of the Kenya Rift Valley, although identification must be uncertain in view of the intervening disturbances.

The age of these peneplains must remain to some extent conjectural in the absence of any evidence of their relations with the marine deposits of the coastal belt, but the presence of Miocene beds overlying the lower peneplain in the Mount Elgon area puts a forward limit to the date of their formation. They may well be much older, for a great amount of erosion of thoroughly consolidated rocks has taken place since their formation.

(b) *Rifting*

It is not relevant here to go into the much-debated question of the cause and mechanics of rift valley formation. It is clear that the rifting was connected with some form of uplift, and that, at least in Uganda, it postdated the most recent peneplanation of the country. It is equally clear that there was more than one period of disturbance in the Albertine Rift; this is evident on several grounds.

(i) The slopes of the main mass of Mount Ruwenzori are well dissected, while those of the northern spur, where the Kisegi Beds occur, show a comparatively immature topography; the fault scarps here are obvious and of fresh appearance. The same difference is apparent between that portion of the escarpment immediately north of Fort Portal and the main south-east boundary of the Albertine basin, the topography of the former being clearly more mature.

(ii) The Muzizi valley and that of the lower part of the Nkussi, both of which are deep and mature, bear witness to the presence of the Albertine depression in some form or other long before the formation of the present bounding fault scarp, over which their streams now plunge abruptly. And both valleys contain near their mouths beds which greatly resemble the Kisegi Beds of the southern end of the basin.

(iii) The actual existence of the Kisegi Beds, which have clearly been greatly affected by later rifting movements, affords evidence of a "Proto-Albert", presumably occupying a basin formed as the result of the earlier phase of movement.

It would appear that Bailey Willis,[1] in his monumental study of the African rift valleys, has perhaps underestimated the importance of erosion as a destroyer of the fault scarps of the older period, and that this has led him to minimise the amount of faulting that has actually occurred. He considers, for example, that there is no clear evidence of faulting along the south-east edge of the Ruwenzori range, but to the present writer it seems most probable that the structure of the main mass of Ruwenzori is comparable with that of its northern prolongation, which consists of a number of fault blocks, heavily tilted towards the valley of the Semliki, and stepped towards the south-east.

[1] *East African Plateaus and Rift Valleys*, Carnegie Inst. of Washington, 1936.

The Pleistocene Succession in Uganda

The writer believes that the great bulk of rift-valley formation was already accomplished by Kaiso, i.e. Lower Pleistocene, times, for he has nowhere observed any very considerable disturbances in the Kaiso Beds, and considers that the deposits found above the scarp in the Muzizi and Nkussi valleys are in no way comparable with the Kaiso series, but are very similar to those of the Kisegi: this point will be more fully dealt with later.

(c) *The Drowning of Uganda, Past and Present*

Perhaps the most striking of all the features in the Uganda landscape is the extent of papyrus swamps and lakes. The drowned topography of most of the shore of Lake Victoria needs no comment, nor does the phenomenon of Lake Kioga. But far more striking are the papyrus-choked valleys of Kigezi and parts of Toro and Ankole, for these are the valleys of ungraded and, in some cases, rapidly flowing streams.

That such papyrus-choking is not a normal state of affairs is clearly shown by the comparative freedom of the Kagera between Kikagati and Kyaka, of the Ruizi below Mbarara and even, at its lower end, of the sluggish Kafu!

Clearly, then, the formation of papyrus swamp is an indication of some form of drowning. Now the only possible causes of drowning in an ungraded stream are:

(a) a blockage, as, for example, by a lava-flow, landslide or fault,

(b) a tilt against the direction of the stream.

The formation of Lake Bunyoni and other lakes in Kigezi is certainly to be ascribed to the former cause, but the remainder of the drowned Kigezi and Ankole valleys can only have been produced by the latter process.

A study of the map reveals that the drowning chiefly affects the upper portions of the north-westerly flowing streams, i.e. those which run into Lakes Albert and Edward. The papyrus swamps extend to the east of the swamp-divides into the valleys of the Kagera and its tributaries. A little lower down these valleys, however, we can find a fairly well-marked head of rejuvenation; the rivers, from being sluggish swamps filling the whole valley, become canalised into definite channels cut into grey swamp deposits of which the latest have been forming right up to the moment of rejuvenation. Clearly, then, the latest tilt has been towards the east or, more probably, the north-east. And such a tilt is, indeed, competent to explain most of the drainage anomalies of Uganda.

The present outlet of Lake Victoria at Jinja may confidently be ascribed to this tilt. The outlet has all the appearance of a simple over-flow, due to the ponding up of the lake until it attained the level of the lowest available col. Such a ponding up might well cause a general rise of level and consequent drowning of the whole of the lake shore; nevertheless, there is considerable contrast between the stretch which lies between Masaka and the Kagera river and the remainder. This stretch presents features of emergence rather than of drowning; for example, the encroachment of papyrus swamp into the lake which has resulted in the cutting off of Lake Nabugabo, and also the substitution in the swamps around Katera and Kiebbe of sword-grass, which indicates only periodical submersion, for papyrus, which needs perennial water. The lower part of the Kagera valley also appears to be recovering from drowning rather than to be sinking further. This is all consistent with the direction of the tilt postulated above.

A further example on a smaller scale of the process which has affected Lake Victoria is provided by the chain of lakes, Nakivali, Kachira and Kijanebalola, through which flows the river Ruizi. Their natural topographical outlet is the broad valley of the Orichinga, which joins the Kagera at Nsongezi, but the present outlet is a narrow channel, another obvious overflow of geologically recent date, at the eastern end of Kijanebalola. It is cut through a low col, like the Ripon Falls, and is, topographically speaking, in almost the last place one would expect to find it. In this case it is particularly clear that only tilting could possibly account for such a reversal.

We see that much of Uganda is at present drowned on account of tilting. Applying the Lyellian method, it at once becomes plain that past drowning accounts for many features at present visible, especially in the Ankole region.

In Ankole there are many flat-bottomed valleys containing no stream or, at most, a small one. Occasional exposures show that they are filled with white or greyish deposits, and Mr Way of the Uganda Geological Survey informed me that it was not unusual to have to penetrate 20 ft.

or more through such deposits in order to reach bedrock. Now it is clear that these are swamp deposits, for exactly similar grey and white clays and silts have been deposited in recent swamps and are now being exposed in some places by river rejuvenation.[1]

It is probable that a certain amount of similar deposition is still proceeding on these flats, for the red washes which cover the slopes do not appear to overlap on to the grey deposits; as the latter are largely argillaceous and hence hold up a good deal of water, the reduction process explained in the footnote is probably at present operative in exactly the same way as under proper swamp conditions, but the initial presence of swamp deposits is the factor responsible for the continuance of the formation of these grey and white beds.

The coincidence of the grey deposits with the flat valley bottoms is most obviously seen in the numerous anthills, which indicate a sharp dividing line between them and the red hillwashes.

Exposures in the old swamp deposits are scarce, as there are very few rivers of sufficient size to have excavated channels through them to any great depth. Only in the Kagera and Ruizi valleys can they be easily examined, and their appearance there will be described below in the stratigraphical portion of this chapter.

These old swamp-filled valleys are to be found all over Ankole and continue eastward into Masaka district, forming a continuous plain descending towards Lake Victoria. Such a plain bounds the Katonga where it is crossed by the Kampala-Masaka road; here it is only from 10 to 20 ft. above present swamp level. Further to the east, the ancient swamp levels are no more to be seen; presumably, they are now at the bottom of the drowned valleys which form the northern edge of the lake. Neither is there anything corresponding to them in the Kavirondo Province of Kenya, for, although local patches of old swamp deposits do occur in places round the Kavirondo Gulf, they are not widespread and their uplift is most likely due to local faulting or tilting.[2]

[1] The grey colour of the swamp deposits is presumably due to the reduction (in standing water by organic matter) of the ferric iron, so obviously abundant in the red hillwashes, to the ferrous condition.

[2] This asymmetrical distribution of the old drowned area with respect to the lake shows clearly that that drowning was due to tilting and not to pluviation, as Wayland has postulated.

We see that the ancient drowning affected the western side of the Victoria basin, in contrast to that of the present day, and it would in fact appear that the latest tilt has resulted in a considerable easterly shift of the lake itself.

There remains the question of the former outlet of the lake. It is, of course, just possible that none existed and that evaporation managed to keep pace with precipitation, but in view of the great westerly extension of the old swamp deposits, it seems probable that the Kagera-Rufua-Berarara through-valley was the overflow channel. It can hardly have been that of the Katonga, for the old swamp deposits do not extend far up that valley, and, in any case, it is inadequate in size to have coped with drainage of such magnitude.

To sum up, it may be stated that the past and present drownings of Uganda provide clear evidence of two tilts, the first towards the west or south-west and the second towards the north-east. The age of the first movement is somewhat doubtful; it is this movement that was responsible for the formation of the 100 ft. ± terrace of the Kagera, but the date of the initiation of this terrace is not firmly established, as no implements and very little fauna have been found in its lower levels.

The second movement was clearly of recent geological date as it post-dated the whole of the 100 ft. terrace, in whose upper part occur Levalloisian and Tumbian tools of fairly advanced types, which belong, in date, to the Upper Palaeolithic of East Africa.

(d) *The Older Terraces*

In many parts of Uganda there occur fragments of terraces, in most cases vestigial, and sometimes associated with gravel deposits of a coarse nature. They have yielded neither fauna nor implements. Examples of these terraces are the 225 ft. ± and 175 ft. ± terraces of the Kafu,[1] the old beach at Kiebbe on the road between the Native Administration Headquarters and the Catholic Mission, the partly gravel-covered erosion platform along the eastern flank of the Simba

---

[1] Implements obtained by Mr Hirst of the Geological Survey from the older Kafu terraces are unrolled, and in view of the rounded character of the gravels in general, it is impossible to believe that the implements can be contemporaneous with the deposition of these gravels.

Hills (both the latter in the neighbourhood of Sango Bay), the boulder-gravel level on the Kampala-Hoima road at Mile 30 and, lastly, the old terrace of the Kagera river preserved near Mile $24\frac{3}{4}$ on the Nsongezi-Kagera Port (Nyakanyasi) road. The latter is the 270 ft. $\pm$ terrace of Wayland (see *R.R.R.E.M.U.* plate XLV). It is probable that similar relics are to be found all over Uganda, and it may eventually be possible to classify them, but it seems dangerous at present to use any such fragmentary evidence in building up a synthesis, however conjectural, for once evidence has been so used, the geological world is apt to take it at more than its face value. On this account, the present writer does not propose to follow Wayland in his attribution of certain of these older terraces to a definite first pluvial period, but would rather leave them in a suspense account until further connecting links are discovered. It is quite likely, however, that some of them are attributable to the wet period responsible for the deposition of the Epi-Kaiso Boulder Beds, to be described later.

(e) *River Development*

The anomalous development of the Uganda rivers has been described by Wayland in various papers. Briefly, its peculiarity is that the east-west and west-east-flowing rivers have no clear watersheds, but occupy through-valleys and are separated only by swamp-divides, which are often very ill-defined.

This phenomenon, together with certain apparently anomalous directions of river flow, such as that of the Kafu, has led Wayland to postulate a wholesale series of river reversals. Now there can be no doubt that certain reversals and diversions have occurred, but some, at least, of the through-valleys seem to be sufficiently accounted for without complete reversal on consideration of the prevalent conditions. We have effectively two different base-levels of erosion, one the Albertine Rift and the other Lake Victoria. The intervening country has been unstable in character, and tilting must have operated now in favour of one base-level, now of the other. What is, then, more natural than that two streams, working back from their mouths along the strike of the softer phyllites of the Karagwe-Ankolean series (which undoubtedly control much of the drainage of Ankole and Kigezi), should meet and

develop an oscillating watershed at their source? Complete reversals
are, indeed, incompetent as an explanation of some of the through-
valleys. At least two tributaries of the Kagera have through-valleys
with affluents of Lake Edward, and it is thus inconceivable that a complete reversal can have occurred by both routes.

With regard to the Katonga and Kafu, it seems to the present writer

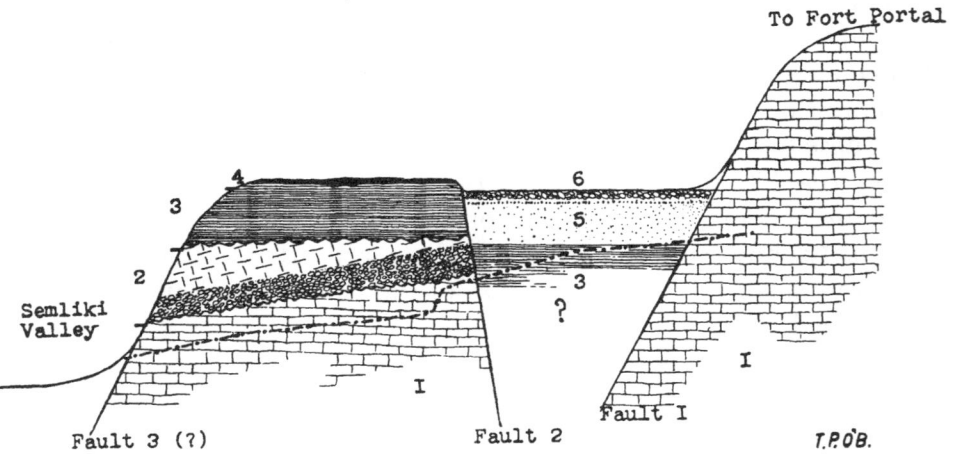

Fig. 1. Cross-section of area between Fort Portal and Semliki Valley, showing
relationship of the lacustrine series to the rift valley faults. Very diagrammatic.
The broken line represents the position of the Kisegi River, approximately, and the
small cliff between the Kaiso and Epi-Kaiso Series opposite Nyambirizi Government
Camp.

6. Peneplain Boulder Bed.
5. Epi-Kaiso sandy clays.
4. Kaiso Bone Bed.
3. Kaiso clays.
2. Kisegi Series.
1. Pre-Cambrian; the area between Fault 3 (?) and Fault 2 represents the most northerly
   extremity of the Ruwenzori upthrust block.

that the valleys through which these reversed streams are supposed to
have flowed, i.e. those of the Mpanga and the upper Nkussi, are much
too insignificant to have drained a large area for any considerable
length of time. Indeed, the upper Nkussi behaves as a hanging stream
to a comparatively minor affluent of the Albertine basin, which has
itself a most impressive valley although its drainage area is very small.

In short, with the possible exception of the present lower Kagera and

Rufua system, the writer does not believe that any of the Uganda rivers have undergone complete reversal.[1]

STRATIGRAPHICAL EVIDENCE

The principal regions where sediments of Pleistocene date are to be found are as follows:

(i) The Albertine basin, around Kaiso, Ndeiga and the valleys of the Muzizi and of the Wasa, Itoja and Kisegi rivers which flow off the northern spur of Ruwenzori.

(ii) The Edward basin, especially at the north-east and east edges of the lake.

(iii) The Kagera valley, especially at Kikagati and around Nsongezi.

(iv) The valley of the Kafu.

(v) Near the Ripon Falls, at Bugungu.

Other small exposures are to be seen in small valley cuttings at several places in the Ankole and Kigezi districts, but they have yielded nothing very significant in the way of evidence.

(i) THE ALBERTINE BASIN

(a) *The valleys of the Kisegi and neighbouring rivers.* The succession in this area may be summarised as follows:

5. "Peneplain Gravel."
4. Epi-Kaiso Beds.
3. Kaiso Beds.
2. Kisegi Beds.
1. Pre-Cambrian (presumably).

1. The pre-Cambrian rocks are hornfelses showing high angles of dip. They need not be further dealt with here.

2. The Kisegi Beds are a series of light-coloured sands and clays, fine-grained except at the base, where a conglomerate is developed, and containing a good deal of selenite in the clay bands. The sandy bands are notably felspathic.

Their unconformable junction with the pre-Cambrian is well shown in the Kisegi valley at the waterfall about half a mile downstream from

[1] It may be noted here that Wayland attributes some reversals to the rise of Lake Victoria owing to pluviation, but it is indeed difficult to see how such a process could possibly reverse more than one river.

25

the point where the telegraph wire crosses the river near Nyambirizi Rest Camp. They differ from the later deposits in their degree of consolidation, which almost entitles them to the name of sandstone. They have also been much affected by earth movements, and show angles of dip of as much as 30°; this also differentiates them sharply from the more recent deposits.

Their thickness is problematical in the absence of detailed mapping, but 600 ft. is hardly an overestimate. They have yielded no organic remains. Their generally fine texture and large felspar content would seem to negative the theory of pluvial conditions during their formation, as has been postulated by Wayland (see *R.R.R.E.M.U.* table facing p. 344).

They are well exposed between Nyambirizi Rest Camp and the Semliki flats as well as in the Kisegi and Itoja valleys (see p. 45, fig. 1).

3. The Kaiso Beds consist for the most part of clays, with a few intercalated sandy horizons which, towards the top of the series, are much indurated with ferruginous matter and contain fragmentary mammalian, fish and reptile remains (see p. 45, figs. 2 and 3).

The clays of the lower part of the series are mainly greenish in colour and contain much selenite, those of the upper part are of the pale greyish type similar to those at Kaiso itself. Their thickness is, in some places, certainly upwards of 200 ft., but the whole of the series is nowhere completely visible. Their dip is nowhere discernible by the naked eye and probably does not exceed 5°. Although they have been affected by small-scale faulting, it is clear that they have not been affected by any major disturbances such as those which have tilted the Kisegi Beds.

They are best exposed in the valley of the Nyabroge and in that of the Kisegi below the waterfall mentioned above; to the south of the Kisegi river they must be banked up against the fault-blocks of pre-Cambrian and Kisegi Beds which make up the northern projection of the Ruwenzori range.

The present writer finds it impossible to consider them as pluvial deposits, for the clays preserve their fine texture up to within a very short distance of the edge of their basin of deposition, which would be quite impossible under open-water conditions with even a normal rainfall. It seems much more reasonable to interpret them as clays formed under swamp conditions, while this would also explain the absence in them of organic remains, which are notably absent in swamp deposits all over Uganda, presumably on account of the decay of bone in the presence of acids formed by rotting vegetable matter.

The ferruginous, sandy, fossiliferous horizons may be interpreted either as (*a*) evidence of slightly more open-water conditions, or (*b*) of more complete desiccation. The former seems to the present writer to be the more satisfactory explanation, in view of the fact that the bulk of the fossil material is of fish, crocodile and hippopotamus, while the rest of the mammalian remains are never found in complete skeletons, as

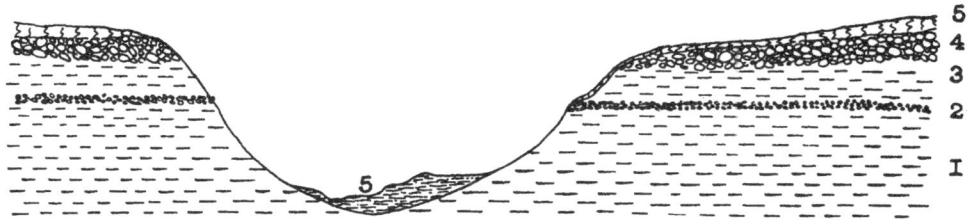

FIG. 2. Cross-section of Nyabroge valley, Toro-Semliki.

    5. Alluvium and hillwashes.
    4. "Peneplain Boulder Bed"; up to 5 ft.
    3. Epi-Kaiso sandy silts; 12–15 ft.
    2. Lower Epi-Kaiso gravel (non-implementiferous); 2–4 ft.
    1. Epi-Kaiso sandy silts; thickness unknown.

they would be if, as suggested by Wayland, they had died around the gradually diminishing pools, but occur here and there in individual pieces apparently washed into their present positions. But in any case, the point is of no great significance, for, as stated above, it is impossible to regard the Kaiso Beds as pluvial in character, and the fossiliferous horizons, therefore, do not necessarily indicate a marked climatic change in either direction.

4. The Epi-Kaiso Beds consist of argillaceous sands and gravels, usually whitish in colour. They overlie and overlap the Kaiso Beds, except to the north-west of a fault which runs approximately north-east to south-west, which appears to have lifted the Kaiso Beds locally above the level of the Epi-Kaiso Lake. This fault appears to die out towards the north-east, where the Epi-Kaiso Beds are again seen to overlie the Kaiso Series.

Their maximum visible thickness is some 100 ft., but their real extent is probably much greater. Except where the slight faulting has occurred as described above, there is no obvious break between them and the underlying deposits, and, as there is no sign of erosion of the Kaiso Beds along the line of faulting before the deposition of the Epi-Kaiso Series, it seems that there is no significant time interval between them.

No implements have been found in the Epi-Kaiso Beds. They are best exposed in the Kisegi valley above the waterfall, in the Itoja valley and in that of the Nyabroge (see p. 45, fig. 4).

5. The "Peneplain Gravel" may possibly be no more than the exceedingly coarse top of the Epi-Kaiso Series, which it overlies and overlaps on to the pre-Cambrian. It consists of coarse, subangular gravel up to as much as 30 ft. in thickness near the edge of its area of deposition, often with a somewhat finer layer in the middle. Its distribution is approximately the same as that of the Epi-Kaiso Beds, but it extends somewhat farther towards the south-west.

Its wide extent and great coarseness would seem to justify the assumption of very wet conditions at the time of its formation and, in the opinion of the present writer, it is one of the very few deposits in Uganda which deserve to be described as "pluvial".

Its upper layers have yielded artifacts of Early Kafuan and Early Oldowan type; these are all of quartz and are considerably rolled. They are described elsewhere by O'Brien.

The Peneplain Gravel forms the surface of a wide plain which slopes down gradually to the north-east. On the surface of this plain Chellean and Acheulean implements have been found. These are all unrolled, though variously weathered, and, in some cases, have been made from boulders derived from the gravel. It follows, therefore, that the true Lower Palaeolithic industries post-date both Kaiso and Epi-Kaiso Beds in Uganda.

(b) *Ndaiga*. There are really no good exposures here, but some small washouts show that Kaiso clays and bone horizons are present, overlain by Epi-Kaiso sands and gravels similar to those described above.

The Kaiso clays are, as usual, fine-grained—although they are exceedingly close to the main Albertine Rift-scarp. They show no signs of any great degree of disturbance, such as would be implied by their participation in any major dislocation along that line.

There is no obvious gravel-covered plain above the Epi-Kaiso Beds here, but a lower erosion platform cut into the Kaiso Beds is overlain by a sheet of very coarse gravel, and it is probable that this corresponds to the "Peneplain Gravel" of the Kisegi area.

(c) *Kaiso.* The upper Kaiso and Epi-Kaiso Beds alone are visible here. The Kaiso Beds of this area have been described by Wayland.[1] Both they and the Epi-Kaiso Beds are similar to those in the Kisegi area, but the Epi-Kaiso Beds are here very much thinner.

The present writer is unable to support Wayland's postulate of a period of desiccation between Kaiso and Epi-Kaiso times on the basis of the evidence available at this locality, for, even accepting the bone beds as evidence of desiccation—and this, as has already been stated, is open to doubt—the thickest bone bed does not, as it should, form the top of the Kaiso Series, but is succeeded by clays of the usual Kaiso type, and it is these latter (pluvial according to Wayland) that are directly overlain by the argillaceous sands and gravels of the Epi-Kaiso Series. Evidently, there is here no real evidence of an important climatic break. It is, of course, possible to make an arbitrary division above the principal bone bed, but the actual change in the type of sedimentation occurs later in the succession.

(d) *Muzizi Valley.* This deep and mature-looking valley flows westwards into Lake Albert plunging over the Rift wall in an impressive waterfall. The part of the valley above the fall is clearly of great age, and contains some deposits of partly consolidated sandstones, which may be easily seen in the neighbourhood of the bridge on the Hoima-Fort Portal road.

These deposits very closely resemble the Kisegi Beds of the Albertine basin, with which the writer is unhesitatingly disposed to correlate them: Wayland's tentative correlation of them with the Kaiso Beds[2] does not bear examination in view of the typical (and quite different) nature of the Kaiso Beds at Ndaiga, which is very near the mouth of the Muzizi, and the fact that while at Ndaiga the Epi-Kaiso gravels succeed

[1] *Occasional Paper No. 2*, Geol. Surv. of Uganda, 1926.
[2] Map No. 1, *op. cit.*

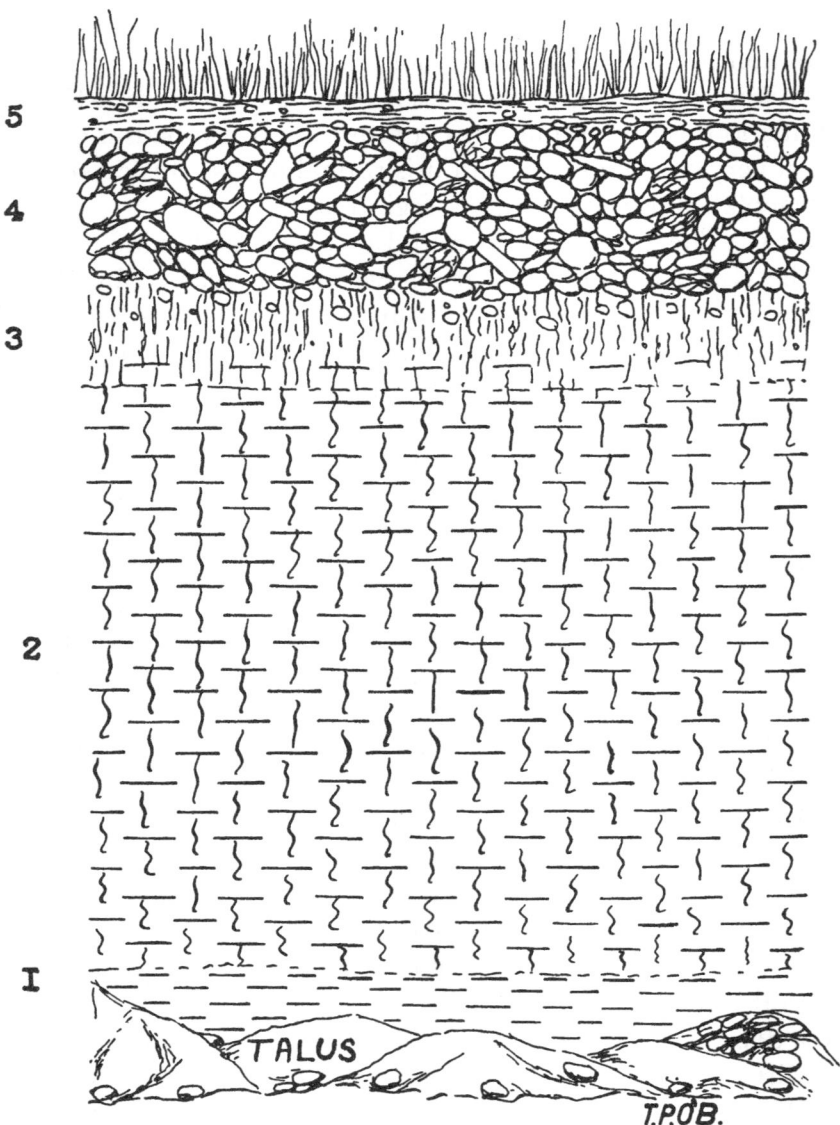

FIG. 3. Muzizi River; 50 ft. terrace gravels overlying Kisegi Sandstone.

     5. Hillwash; 6 in.
     4. 50 ft. terrace gravels; 2–3 ft.
     3. Reddened upper part of weathered zone; 1 ft. 6 in.
     2. Weathered zone of Muzizi Sandstone; 8 ft.
     1. Muzizi Sandstone, pale yellow, very fine.

the Kaiso Beds without a break, there is no corresponding wide spread of gravel in the Muzizi valley, as would be expected if Wayland's correlation were valid.

The only gravel present in the Muzizi valley consists of torrential gravel filling small lateral valleys and subsequently left as a small terrace of ungraded nature as the result of further erosion. On account of its height above the river at the bridge, we refer to this gravel as the 50 ft. ± terrace. The nature, both of the gravel and the rolled implements found in it, support its correlation with the "Peneplain Gravel" of the Kisegi area.

(ii) THE EDWARD BASIN

Kaiso and Epi-Kaiso Beds occur here, the former being fossiliferous as in the Albertine basin. The general aspect of the beds is so similar to those of the Albertine basin as to suggest that both depressions may have been connected at that date. It has been stated that the Kaiso and Epi-Kaiso Beds pass laterally into the volcanic deposits of the Katwe district, but both at Mwaya, at the mouth of the Kazinga Channel, and at Mile 99 on the Fort Portal-Mbarara road, where the Epi-Kaiso Beds are overlain by volcanic ash from the Katwe and neighbouring explosion vents, the beds themselves were of the usual, quartzose, gravelly type and, apparently, contained no ash whatsoever. This is also true of Epi-Kaiso deposits seen at Ruchuru, in the Belgian Congo, at the southern end of the basin, where they appear to pass under the great mass of volcanic materials erupted by the Bufumbiro volcanoes. And, indeed, in view of the obviously recent date of the vulcanicity in this area, it would be surprising if it should prove to have persisted from as early a date as the Lower Pleistocene.

(iii) NSONGEZI AND THE KAGERA VALLEY

Nsongezi is the point at which the Kagera river emerges from a comparatively narrow gorge into the broad and mature valley which carries a minor tributary, the Orichinga, and which continues right down to the broad flats round Lake Victoria.

It is of exceptional interest as being the only important locality in Uganda where stratified implementiferous sediments are exposed. These deposits are for the most part the products of the drowning which affected the whole of Ankole, and Wayland has described them as the "100 ft. terrace" deposits of the Kagera valley, but it must be emphasised that they do not mark a pause in the degradation of the valley after the manner of ordinary terrace deposits; the valley was excavated to something like its present level before their formation, and the terrace itself is a purely aggradational one.

The generalised succession in the 100 ft. terrace deposits and natural exposures in the Orichinga valley near its junction with the Kagera, is as follows:

9. Grey and white clays and silts; 6 ft.±
8. Impersistent thin sandy layer with rubble (O); sometimes apparently represented by a reddening in the clays. The rubble, where present, contains much chipped vein quartz, but no definite tools were found in it by us; 1 ft. ±
7. Grey and white clays and silts; 15 ft.±
6. Sandy bed with rubble, sometimes implementiferous (N), but without much chipped vein quartz; 1–2 ft. ±
(Erosional unconformity)
5. Pale silts with sandy layers; 4 ft. ±
4. False-bedded sands with implementiferous horizons (Levalloisian) in some places; 15 ft. ±
3. Hard, indurated implementiferous rubble (M); 1 ft. ±
2. Sands and clays; 50 ft. ±
1. Boulder Bed; 2 ft. ±[1]

Thicknesses are given as an indication of the order of magnitude; they are very variable.

1. The boulder bed at the base may, perhaps, be evidence of a torrential phase, but it would not be difficult to imagine the formation of a similar bed at the bottom of the valley under present-day conditions.

2. The succeeding sands and clays have an open-water appearance; there are no notably coarse bands among them, and they are probably lacustrine. Some of the finer layers have the grey or white colour associated with swamp deposits.

[1] The Geological Survey has proved a depth of 12 ft. for this boulder bed at Nsongezi. *A.R.G.S.U.* 1936, p. 14.

3. The M-Horizon has been described by Wayland. There does not appear to be any evidence of erosion of the underlying beds before its deposition; its formation is, therefore, to be ascribed purely to desiccation. Such a drying up of the arm of the lake which occupied the Kagera valley would leave a sluggish stream occupying the valley floor, around which prehistoric man would of necessity congregate: this explains the enormous number of implements found in this horizon.

Examination of the tools from the M-Horizon has revealed that the rolled examples are consistently of an earlier type than the fresh ones (Phase B of O'Brien). And as the great bulk of specimens from any one pit are uniformly either rolled or unrolled, it appears that the M-Horizon may have two substages, of which the earlier is more in the nature of a gravel and the later more of a rubble. The present writer was not able to obtain definite confirmation of the presence of the two stages, as no fresh excavation in the M-Horizon was undertaken while he was at Nsongezi. The rolled implements of the M-Horizon (O'Brien's Phase A) may well correspond in age with those found in some of the various beaches, remnants of which exist at many places in the Kagera valley, and which are described below.

4, 5. The Levalloisian-bearing sands and the overlying silts, which are of the usual swamp type, require no special description. They are, in places, almost completely cut out by the erosion which took place before the deposition of the N-Horizon Rubble.

6. The N-Rubble is a fairly persistent horizon. It is not indurated like the M-Horizon, and yields Proto-Tumbian and Levalloisian tools— the latter are not known from the M-Horizon. A little way down the valley below Nsongezi (about a mile from the Tsetse-fly Smokehouse on the Nyakanyassi road) this horizon may be seen directly overlying the phyllites which form the bedrock in this valley.

7. These are typical swamp deposits, mostly very fine. They have yielded a few scattered implements.

8. This horizon is similar to N, but far less persistent. It cannot be traced far up the Orichinga valley, but occurs in Wayland's big pit at Nsongezi and in the exposures near the Orichinga-Kagera junction. It appeared at one time that it was the horizon from which the more advanced Tumbian tools were washing out, but none could be obtained

in situ, and it seems that they occur throughout the upper part of the terrace deposits (beds 7–9).

9. These are similar to 7.

The red valley-side hillwashes do not appear to overlap on to the terrace deposits, and are probably, for the most part, contemporary with them, but there is a widespread brown, loamy surface soil which is obviously later than the latter, and which has yielded some Late Levalloisian or Still Bay.

It must be emphasised that there is nothing in the upper deposits of the 100 ft. terrace which in any way suggests pluvial conditions. The deposits are, for the most part, very fine-grained in character right up to the outside edge of the terrace: this would be quite inconceivable under conditions of heavy precipitation. It seems impossible, therefore, to attribute the formation of these beds to a rise in the level of Lake Victoria owing to an increase in rainfall, and the alternative hypothesis of a regional tilt to the west or south-west must, therefore, be adopted.

The 30 ft. ± Terrace

There is a rock platform below the 100 ft. ± terrace at Nsongezi which is covered here and there by a layer of gravel and coarse shingle, up to 15 ft. thick in some places. It does not appear to be of any great significance, but it is noteworthy that the only implements obtained from it are of M-Horizon (Middle Acheulean) type. O'Brien was inclined, on this account, to refer its formation to some period between the deposition of the M-Horizon Rubble and the Levalloisian-bearing sands, and to consider the present terrace as re-excavated. This is, however, inherently improbable in the absence of any other evidence of post-M-Horizon erosion,[1] and it is much more likely that the gravel consists almost entirely of material from the basal boulder-bed of the 100 ft. terrace, and the overlying shingly horizons redeposited more or less *in situ*. The extraordinary abundance of implements in the M-Horizon and the friable nature of much of the Levalloisian and Tum-

---

[1] Dr Solomon has, however, overlooked the existence of some scattered patches of boulder gravels in the bottom of the Orichinga in which, again, the only tools found by us were rolled M-Horizon types. I cannot agree with his explanation of the latter fact, for the great bulk of Proto-Tumbian and Tumbian implements of the post-M beds were usually large and made in good, durable quartzite. (T. P. O'B.)

bian material are quite sufficient to explain the exclusive occurrence of the former in the 30 ft. terrace.

There are similar low-terrace gravels at various places lower down the Kagera; at Nyakanyassi, such gravels cover a bench cut in the 100 ft. terrace silts. The coarseness of these gravels would indeed seem to indicate pluvial conditions, since the large boulders are some distance from any possible source (see p. 47, fig. 4).

None of the 30 ft. gravels are products of a graded stream; they are entirely composed of coarser material; material of sand grade, which is characteristic of ordinary graded terrace gravels, is here completely absent.[1]

Old Beaches in the Kagera Valley

At various places along the valley below Nsongezi and in the Orichinga valley a good deal of water-rolled material is to be found well above the level of the top of the 100 ft. terrace. It has mostly been incorporated, with much unrolled material, in later hillside rubbles; indeed, in many borrow pits opened to obtain material for the Nsongezi-Kagera Port road, it may be seen that the rubble exposed in them is entirely free from waterworn components except in the topmost 5 or 6 ft.; the occurrence of such material is thus, in some instances, a useful indication of rubbling of more than one age.

The most obvious of these beaches is that exposed at Mile 14 on the Nsongezi-Mbarara road, but there are also exposures at the base of the hills behind Nsongezi P.W.D. camp and at many places along the Kagera Port road, notably at Mile $24\frac{3}{4}$ (see p. 47, fig. 2).

These beach deposits are often associated with small local benches which do not appear to be constant in height and may, perhaps, be representative of several ancient lake levels. They are, in fact, similar to the kind of small and ephemeral beaches now being formed around the drowned shores of Lake Victoria, to which they probably corresponded.

Rolled implements have been found occasionally in the deposits, most of them Chellean, but the beach at Mile 14 appears to represent one of the latest examples of the series, as the waterworn implements

---

[1] At Nsongezi, however, the 30 ft. terrace is composed largely of sand and shingle with only occasional boulders.   (T. P. O'B.)

found in it are approximately of the stage found rolled in the gravelly facies (O'Brien's Phase A) of the M-Horizon, i.e. Early-Middle Acheulean (see p. 49, fig. 3). Other of the ancient beaches may probably correspond to the lower part of the 100 ft. terrace deposits.

At Mile 24¾ a large patch of beach gravel is preserved. In these gravels, which represent Wayland's 220 ft. ± terrace, rolled implements of Kafuan type occur. It may be that the coarse gravels of Wayland's

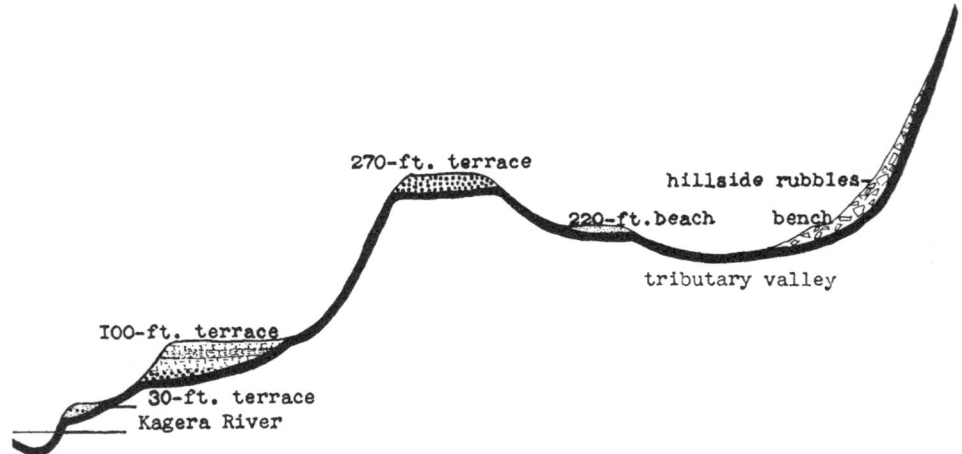

Fig. 4. Diagrammatic cross-section of Kagera valley about 15 miles downstream of Nsongezi. No scale, but the vertical distances are approximately in correct proportions.

270 ft. ± terrace (see p. 47, fig. 1), also preserved close by, which contain no implements, were rearranged at a level some 50 ft. lower, presumably during the course of the formation of the 220 ft. beach.

The Kikagati Gravels

The writer was fortunate enough to be on the spot when a large excavation, some 60 ft. deep, was being made in old river gravels at this point, about 8 miles above Nsongezi, for the purpose of erecting a hydro-electric plant (see p. 49, fig. 1).

The section disclosed a series of sands and gravels, having the appearance of ordinary deposits of a graded river. The lowest levels were somewhat indurated with calcareous matter, and from them the excavators had recovered some fossil elephant teeth and had luckily preserved them. They have been identified by Dr Hopwood (see Appendix B) and he regards them as probably of Lower Pleistocene date, although, in view of the small amount of material of this age available from African sources, he does not wish to be too positive on this point.

The middle part of the section is mostly in false-bedded sands, the top 6 ft. are again gravelly, but contain much earthy, ferruginous material and, in fact, appear to have been rearranged subsequently to the formation of the bulk of the terrace.

No implements were found in the principal section by us, but Mr Wayland showed me two specimens of "Oldowan" type from here. But in shallow pits in the topmost gravel there occurred some Acheulean implements of the type associated with the M-Horizon at Nsongezi, and, in view of this fact, and of the absence of the ordinary upper part of the 100 ft. terrace swamp deposits over the area where the implementiferous gravel occurs, it would appear that the latter may be of the age of the 30 ft. terrace. It would appear that the bulk of the sands and gravels must be correlated with the lower part of the 100 ft. terrace at Nsongezi, for they are the deposits of a graded river, and such a river must have had its base-level in the Victoria basin; the alternative, Lake Edward, is far too low in level, and the great intervening valley of the Berarara, which forms the "complement" to that of the Kagera, must have been cut long before Pleistocene times. The somewhat coarser nature of the Kikagati deposits compared with those at Nsongezi is explained by their proximity to the narrows at that place.

The M-Horizon was not represented in the hydro-electric section, but it is quite likely to be present under the comparatively recent swamp deposits that fill the wide valley above Kikagati.

The Ruizi River at Mbarara

The section by the bridge over the Ruizi shows quartzose gravel running under grey swamp clays which are being worked for brickmaking. The writer has seen no implements from this locality, and in the absence of definite horizons no detailed correlation is possible, though the deposits clearly correspond to some part of the 100 ft. ± terrace of the Kagera.

(iv) The Valley of the Kafu

Wayland and Hirst have investigated the Kafu terraces in far greater detail than the present writer could possibly do, and the details of their work are to be found in the *Annual Reports of the Geological Survey*, 1926 and 1927.

Only the 50 ft. ± and flats terraces were studied by the present expedition. Excavations were made in both terraces at Kinoga, about 6 miles below the bridge on the Kampala-Hoima road, and at the ferry on the direct Masindi-Kampala road, about another 35 miles downstream.

The 50 ft. terrace is of doubtful nature; it is difficult to regard it as an ordinary graded river terrace, as it consists entirely of pebbles; finer material, except for a little interstitial clay which has, no doubt, percolated from above, is completely absent. It looks, in fact, like a shingle beach.

According to the observations of Hirst (*A.R.G.S.U.* 1926, p. 15) the 50 ft. terrace shows a marked tilt in the present downstream direction of the Kafu and actually disappears below the flats terrace. Now, as the latter is not more than a quarter of a mile wide at the ferry on the Masindi-Kampala road, it follows that the 50 ft. terrace here and farther to the east must be still narrower and cannot, therefore, be attributed to a ponding back of Lake Kioga. It seems reasonable to consider it as the beach of a local lake, formed as the result of a regional westerly, or south-westerly, tilt fed by a reversed Kafu from the Kioga drainage area, and with some kind of overflow towards Lake Albert.

On this hypothesis the Kafu already existed in approximately its present form before the formation of the 50 ft. terrace, and has not been greatly eroded since that period. Some of the subflats material (below present swamp-level) may possibly be considered as lacustrine deposits corresponding to the 50 ft. terrace (or, as the present writer would prefer to call it, the 50 ft. beach).

A subsequent easterly tilt would naturally result in a reversal of the drainage, with the formation of the flats terrace and, eventually, in the lower reaches of the stream, in rejuvenation causing the river to cut into the flats terrace gravels. Such rejuvenation is well seen at the ferry, where the river occupies a narrow channel cut into gravels which, on account of their nature and contained implements, must be referred to as the flats terrace. The head of rejuvenation is somewhere between the ferry and Kinoga, where the river still flows in a broad swamp little below the flats terrace. The gravels of the latter contain typically rather more interstitial material than the 50 ft. gravels, and, while this material tends to be somewhat more argillaceous in nature than the nature of an ordinary graded terrace gravel, it would be consistent with formation in and around a swamp.

If we correlate the Kafu terraces with those of the Kagera on the basis of the tilts postulated above, it follows that the 50 ft. beach covers the period up to the end of 100 ft. terrace times, while the flats terrace entirely post-dates it. It is possible, however, that the 50 ft. terrace lake may not have persisted for any great length of time, for, if the outlet towards Lake Albert happened to erode through fairly soft material, such a lake would be drained very rapidly. Thus, the 50 ft. terrace may correspond only to the earlier stage of the 100 ft. terrace of the Kagera.

Such a correlation[1] is supported by the similarity of the implements from the 50 ft. beach with those of the Kagera 220 ft. ± beach at Mile $24\frac{3}{4}$ on the Nsongezi-Kagera Port road.

(v) BUGUNGU, NEAR RIPON FALLS, JINJA

Around Bugungu, to the west of the Ripon Falls, various shallow sections and excavations have exposed rubbles, laterites and boulder beds. Some of them have yielded implements, but the number of exposures does not appear to be sufficient to establish a definite climatic sequence, and Wayland's tentative correlation of one of the rubbles with the M-Horizon at Nsongezi seems to be at variance with the implements found in it.[2]

---

[1] A similar correlation was worked out by O'Brien independently, but on the same evidence. His interpretation, however, differs from mine in attributing the 50 ft. gravels to a direct reversal of the Kafu, following the south-westerly tilt, while admitting their beach-like appearance.

[2] See Plate XLIX, fig. 2, *R.R.R.E.M.U.* The M-Horizon is of Middle Acheulean date, but in this rubble we only obtained some rare examples of Upper Acheulean type, the rest of the material being variously Proto-Tumbian and Levalloisian of Upper Palaeolithic date. (T. P. O'B.)

The lateritic deposit above Bugungu from which Late Kafuan tools were obtained is of somewhat greater interest. It has the appearance of stratification and may possibly be water-deposited—although the point is not an easy one to decide in the case of deposits of this nature. If, however, this is the case, it probably belongs to a high-level stage of Lake Victoria immediately prior to the cutting of the Ripon Falls, and as a result of the last, north-easterly tilt. For this suggestion I am indebted to O'Brien.

THE PLUVIAL HYPOTHESIS

Mr E. J. Wayland's synthesis of the Uganda evidence is based on the supposed occurrence of a number of pluvial periods during the Pleistocene, and in the past both Dr L. S. B. Leakey and the present writer, working in Kenya and Tanganyika, have followed him in interpreting depositional phenomena in those countries on a similar basis.

The present writer now feels compelled to regard this basis as extremely insecure. His examination of the Uganda evidence has led him to the conclusion that many of the phenomena ascribed by Wayland to climatic agencies are far more easily and logically explained by earth movements, and that in some cases, indeed, there are definite indications of non-pluvial conditions in terrace deposits whose formation is ascribed by Wayland to a rainy epoch.

Only two pieces of the Uganda evidence appear to the writer to indicate a precipitation greater than that at the present day. These are (a) the coarse nature of the Epi-Kaiso "Peneplain Gravel" and (b) the coarse gravel of the 30 ft. terrace of the Kagera at Nyakanyassi. All the other gravels and boulder beds might easily be formed by streams similar in size to those of the present day.

In the light of the insufficiency of the pluvial data in Uganda, it becomes necessary to re-examine the Kenya and Tanganyika evidence. Most of that evidence was obtained in the closed basin of Nakuru and Naivasha and is, in any case, concerned with "Gamblian", i.e. very recent times. It now seems quite likely that the possibility of post-Gamblian movements was somewhat underestimated in the early work of Dr Leakey and the writer, and while it is clear that Lakes Nakuru and Naivasha formerly covered a wider area than at present, they may not have attained the height above their present level that the distribution of their sediments has previously been taken to indicate.[1]

A comparatively small change in the rainfall is competent to produce a great alteration in the area of these lakes, as is shown by the recession of Lake Naivasha during the present century. The writer, therefore, considers that while the former extent of these lakes undoubtedly indicates an increased precipitation, it need not have been of dimensions corresponding, say, to a glacial period in the northern hemisphere.

The question of the pluvial nature of the Kamasian Beds is more complex. There are undoubtedly considerable boulder gravels within this series which may indicate pluvial conditions, but the old "Lake Kamasia" must have owed its existence primarily to topographical rather than climatic factors. In any case, the Kamasian, as we know it to-day, probably embraces sediments of all kinds of ages from Miocene onwards. No detailed work has been done on the deposits, and in its absence it is not justifiable to make any inferences as to their character and age.

The evidence of the Oldoway Beds is not especially indicative of pluvial conditions; the fauna contains no elements of notably "wet-loving" type and, if not extinct, might be found living to-day, under distinctly "interpluvial" conditions, in such a locality as Baringo. Moreover, the mineral content of the deposits provides definite indications of non-pluvial climate, especially in Beds I–III, which are entirely volcanic ash and contain practically no detrital material such as must be contributed by streams flowing over the surrounding pre-Cambrian terrain. Only in Bed IV is there any appreciable quantity of such an essential sedimentary mineral as quartz! And, strangely enough, it is just that bed, with its comparative abundance of antelope remains, that has hitherto been considered as less pluvial in character than the rest of the series.

All these considerations go to show that the Pluvial Hypothesis rests on very slender foundations, and the writer is inclined to discard it completely as a basis for the classification of the African Quaternary.

---

[1] The writer has never been able to follow Dr Nilsson's work on the ancient beaches around those lakes; their identification has always seemed to him to be speculative and of doubtful validity.

In his opinion, it will have to be replaced by palaeontological, archaeological and geomorphological methods. He is further of the opinion
that there is no reliable material for the institution of Glacial-Pluvial
correlations or for meteorological theories of the Ice Age which are, in
any way, dependent on such conditions.

The writer does not wish dogmatically to assert that no such correlations will ever be possible, but he is emphatically of the opinion that the
present evidence is absolutely insufficient for their establishment.

SUMMARY AND CONCLUSIONS

The sequence of events in Uganda is interpreted by the writer as
follows (see also Table II, after page 316):

(a) *Pre-Pleistocene* (but included in the Pleistocene by E. J. Wayland).

Deposition of the Kisegi Beds and subsequent formation of the later fault-
scarps of the Albertine Rift, including the main south-eastern scarp.

Formation of the 225 and 175 ft. terraces of the Kafu and of the 270 ft. terrace
of the Kagera.

Little or no evidence as to climatic conditions in immediate pre-Pleistocene
times.

(b) *Lower Pleistocene* (palaeontological division following A. T. Hopwood).

Deposition of Kaiso Beds under swamp conditions in the Albert and Edward
basins. The similarity of the deposits in the two basins suggests their continuity.

The base of the gravels at Kikagati, Kagera valley, may belong to this epoch.

The position of the Epi-Kaiso Beds is doubtful; the implements found in their
upper part suggest a correlation with the lower beds at Oldoway which are
Middle Pleistocene, but it is possible that these primitive industries (Kafuan and
Oldowan) were already in existence in Lower Pleistocene times. The same
remarks apply to the implementiferous gravels of the Muzizi valley (50 ft.
terrace).

The nature of the gravels at Kikagati, which are normal deposits of a graded
river, indicates that the Kagera must, at that period, have been graded to Lake
Victoria, which is the only possible local base-level; Lake Edward lies too low
to have given rise to graded deposits as far upstream as Kikagati.

There is not much evidence as to the climate at this period, except that the
Epi-Kaiso Peneplain and Muzizi gravels indicate wet conditions at its close.

(c) *Middle Pleistocene.*

Part of the 100 ft. terrace of the Kagera must be attributed to this era; the
lowest portion of all probably corresponds to the basal gravels at Kikagati,

which may be Lower Pleistocene, but the M-Horizon appears to correspond approximately to part of Bed IV at Oldoway, judging by the implement types and, as there is no evidence of much unconformity between the M-Horizon and the underlying strata, it seems probable that some, at least, of the latter must come within the Middle Pleistocene. The drowning of Ankole must, therefore, have begun at this time, and is clearly indicative of a regional south-westerly tilt.

The sporadic remnants of beaches in the Kagera valley which contain Lower Palaeolithic implements also belong to this period. It is probable that at some time during the period the outlet of the Victoria basin was through the Kagera valley into Lake Edward.

There is no evidence of pluvial conditions during this period, but the M-Horizon provides proof of aridity near its close.

(d) *Upper Pleistocene.*

The 100 ft. terrace of the Kagera above the M-Horizon all belongs to this division, for Levalloisian implements occur more or less throughout, and they have nowhere in East Africa been recorded in deposits of earlier date.

During the first part of this period the valleys of Ankole were entirely swamp-filled and were more or less inlets of the contemporary Lake Victoria. Conditions do not appear to have been notably wet.

A regional, north-easterly tilt then resulted in the reversal of a large part of the courses of many of the rivers, and in an easterly or north-easterly shift of Lake Victoria causing the drowning of its shores in that direction and the establishment of a new outlet at Jinja.

The flatter portions of the westward-flowing rivers were drowned and filled with papyrus swamp, while the eastward-flowing streams were rejuvenated, and cut through their swamp deposits of the preceding periods. This process is still continuing.

During this period of rejuvenation there seems to have been a period of somewhat abnormal rainfall, which gave rise to the torrential 30 ft. terrace of the Kagera. This may correspond to part of the Gamblian of Kenya.

THE AGE OF THE PALAEOLITHIC CULTURES

The precise age of the Kafuan and Oldowan cultures is still doubtful. It seems certain that these primitive types persisted in some places into Upper Palaeolithic times[1] or even later, but their first occurrence, in the top of the Epi-Kaiso Beds, is clearly older than that of the true Lower Palaeolithic cultures, beginning with the Chellean.

[1] There is no evidence of the persistence of Oldowan culture after Early-Middle Acheulean times.

PLATE I

Fig. 1 : top left. Base of the Kisegi Series resting on pre-Cambrian rocks. Exposure in the Kisegi valley, Toro-Semliki area, Albert Rift Valley.

Fig. 2 : top right. Upper part of the Kaiso Series exposed in the Kisegi valley. Patches of indurated bone and shell bed occur on the surface.

Fig. 3 : bottom left. One of the Kaiso indurated bone and shell horizons. Exposure at Kazinga Channel, Lake Edward.

Fig. 4 : bottom right. Upper part of the Epi-Kaiso Series, just east of their unconformable junction with the Kaiso Series, near the Kisegi valley. The Epi-Kaiso "Peneplain" Gravels are not present here but the lower, non-implementiferous gravel horizon occurs near the top of the section.

PLATE II

Fig. 1: top left. The Kagera 270 ft. ± terrace seen from the level of the 100 ft. ± terrace plain. Near Mile 24¾ on the Nsongezi-Kagera Port Road.

Fig. 2: top right. The Kagera 220 ft. ± beach gravels at Mile 24¾.

Fig. 3: bottom left: The upper part of the Kagera 100 ft. ± terrace deposits near Nsongezi. These clays produced Levalloisian and Tumbian industries. The figure stands on the upper part of the white sandy N-Horizon Bed at the base of the clays.

Fig. 4: bottom right. Kagera 30 ft. ± terrace gravels resting on a bench cut into 100 ft. ± terrace silts. Exposure at Nyakanyassi.

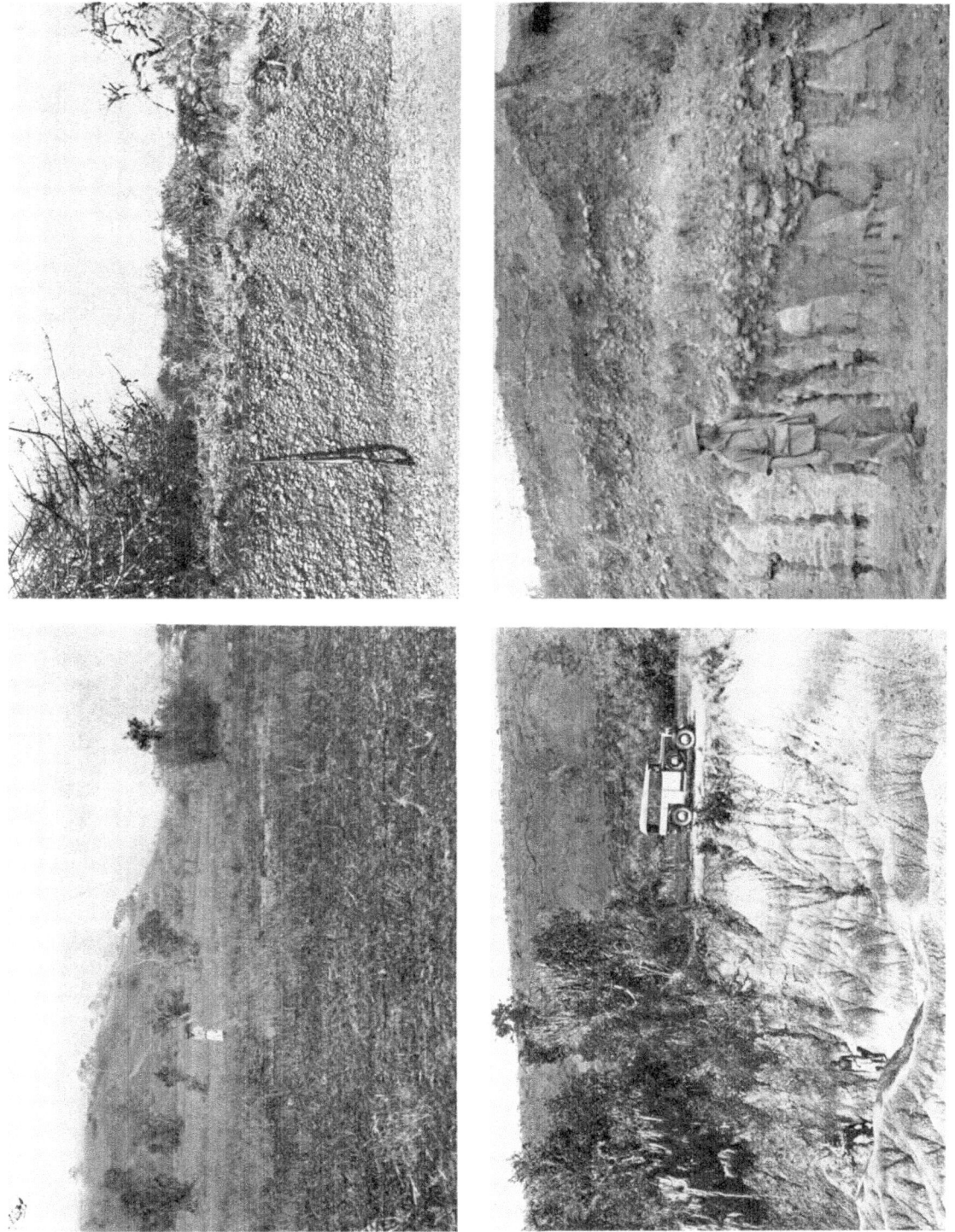

PLATE III

Fig. 1: top left. Basal gravels of the Kagera 100 ft. ± terrace resting on "fossil" granite cascades. Section exposed during the construction of the Mwirasandu Mine Hydro-electric Station at the cascades near Kikagati.

Fig. 2: top right. The Younger Rubble. Exposure at Kikagati.

Fig. 3: bottom left. The Early-Middle Acheulean beach gravels near Nsongezi.

Fig. 4: bottom right. A view of the Kagera valley looking upstream. In the right middle distance is seen the edge of the Pliocene erosion platform, beyond (some 30 ft. lower) is a small shelf belonging to one of the early Kagera terraces, perhaps the 270 ft. The tree-covered plain is mainly occupied by 100 ft. ± terrace deposits.

Unfortunately no fauna has yet been obtained from these beds, and in its absence it is impossible to ascribe them to any division of the Pleistocene, though in view of their somewhat intimate connection with the Kaiso Beds of undoubted Lower Pleistocene date, the writer is inclined to refer them to the same epoch.[1] Mr Wayland has shown him an artifact of apparently Oldowan type from the basal gravel at Kikagati, which also yielded the fossil teeth of an elephant which occurs at Kaiso in the Bone Beds. Apart from this, however, there is no definite evidence of the presence of Palaeolithic man before the Middle Pleistocene.

The Chellean and Acheulean cultures are known from the Oldoway evidence to be of Middle Pleistocene date, and no further data have been forthcoming in Uganda.

The Tumbian is clearly contemporary with much of the Uganda Levalloisian and must, therefore, be referred to the Upper Pleistocene.

In conclusion, the writer wishes to express his thanks to Mr Wayland for some stimulating conversations and to Mr O'Brien for suggestions in the field and elsewhere.

This work was accomplished with the aid of a Senior Studentship of the Royal Commission of 1851.

[1] Personally, I prefer to regard the Early Kafuan and Early Oldowan of the "Peneplain" and Muzizi gravels as belonging to the beginning of the Middle Pleistocene. (T. P. O'B.)

CHAPTER IV

Terminology and Cultural Interrelations

IN describing the Stone Age cultures, I have followed established terminology wherever it seemed applicable, with due regard to typology and, particularly, technique. Recently, the use of European terminology for foreign cultures has come in for some criticism, on the grounds that it should be reserved for European cultures, because mere similarity is not proof of a family tie. This may be true, if only typological likenesses are considered, but typology is frequently a secondary characteristic, imposed by the use of some special technique, and so it is necessary to seek whether technical causes have induced the typological similarities, before condemning the latter as valueless for tracing cultural relationships.

Naturally, this applies more particularly to the more complex cultures and their subdivisions. A simple core and flake industry of Clactonian type may be so unspecialised in its technique as to be difficult to relate to the classical European Clactonian. In other words, though the typology of such an industry may appear Clactonian, we cannot, necessarily, regard it as a branch of that culture simply on account of *retroussé* platforms and prominent bulbs. All we can say is that the extremely simple Clactonian type of technique was employed.

In cultures as complex as the so-called hand-axe groups and the Levalloisian, on the other hand, technique played an important part in producing *form*—that is, typology. In such cases, therefore, technique and typology are almost synonymous and, unless we believe in the independent invention and development of these complex techniques, we cannot, in the present state of prehistory, do otherwise than regard close similarities between certain industries as due to the spread of technical methods. Thus, it seems logical to regard such industries as members of single, expanded cultural entities, without, of course, demanding their contemporaneity.

At the same time, it is most necessary to recognise the part played by environment and varying raw materials in modifying local industries, but, again, such industrial differences must not be so magnified as to be made to assume the importance of *cultural* characters which would mislead us into giving new names where they were not required. In fact, the prehistorian's task, in this respect, is to try to evaluate the proportion of already known cultural characters in given groups and to see whether these are cancelled out by the local industrial differences imposed by different environments, different requirements and, above all, different kinds of stone. If they are not, we may consider the local group to be a branch or colony of the mother-culture, though possibly very much modified.

Thus, I have no hesitation in calling our Uganda tortoise-core and facetted-flake culture Levalloisian, and, by so doing, implying some connection with the prototype. In my opinion, that prototype was European in its origins and earlier history, but it may yet be proved that the European Levalloisian was itself only a branch of a culture whose roots lay elsewhere.

In the following pages, therefore, cultures which closely resemble those already recognised elsewhere have been called by those names, but, in all cases, the generic title of Chellean, Acheulean, etc., should be considered as being prefixed by the adjective *African*, though I have omitted doing so in the text, for the sake of brevity. A culture which appears to be unlike any other already known, or too much modified for original characters to have been preserved, will receive an appropriate new name.

Leakey's discovery of the Lower Palaeolithic series at Oldoway was of prime importance to the study of East African cultures, for it seemed to provide, for the first time, a perfect standard succession by which to evaluate the many isolated Lower Palaeolithic industries belonging to the Chelleo-Acheulean group. Further, it seemed that Oldoway could provide the clearest possible evidence of technical and typological changes which could be considered of regional significance and as a basis for broad correlations with hand-axe cultures beyond East Africa.

Unfortunately, the Oldoway succession has not yet been published, so we have not the results of Leakey's detailed study available for comparative work. However, there are several small collections of material from Oldoway, notably that in the Coryndon Museum, Nairobi, and this series is classified as follows:

Bed IV. Acheulean stages 2–6.

Bed III. Really the upper, reddened part of Bed II. Contains a single industry, Early Acheulean, stage 1.

Bed II. Chellean stages 1–5, from the base of Bed II to the base of Bed III.

Bed I. Oldowan core-choppers; pre-Chellean.

Wayland, who had visited Oldoway, had suggested a correlation between Bed III, with its Early Acheulean industry, and his M-Horizon stone-bed of the 100 ft. $\pm$ terrace of the Kagera, in Uganda.[1] Apart from this, however, no other correlation had been effected or, perhaps, was possible, between the Oldoway industries and those of Uganda, and it was not until after we had left the country that we were able to come to any significant conclusions about correlations with the Oldoway series. We had, however, realised that the M-Horizon material was not Early Acheulean but a later stage, and so could not be the same age as Oldoway Bed III. Later comparative study showed that the M-Horizon Acheulean was, in fact, more akin to Acheulean stage 4, at Oldoway, though there were important differences.

This industry was not the only one to differ in some respects from the classical Oldoway groups, for we soon realised that almost all our culture stages were greater or lesser departures from the Oldoway "norms". The question then arose, were the latter anything more than *local* norms? We saw how different kinds of rocks considerably modified Acheulean technique, and that the absence of all but small quantities of fine-grained materials even caused the makers of certain cultures to shun Uganda, though next door, in Kenya, they flourished exceedingly.

It became more and more evident that we should have to treat some of the Uganda industries as localised variations, not necessarily involving different racial or cultural entities, but technical adaptations, imposed by local conditions upon the flexible, African mother-culture.

There is nothing very original about this conclusion, and it has long ago been recognised by South African workers, but it needs to be

[1] *R.R.R.E.M.U.*

53

emphasised in connection with East African work, otherwise there will be danger of "standard sections" such as Oldoway becoming the law of the Medes and the Persians over too wide an area.

For these reasons, then, it is not easy to effect close correlations between the various East African industries, though broader linkage between the cultures can easily be recognised. We must rely less upon the culture-stages and more upon geology for providing correlative evidence, and even this is sometimes a tricky guide. If all East African Pleistocene deposits were as well displayed as at Oldoway, close correlations would probably be simple, but, as it is, there are large gaps and important anomalies in various places which demand much more detailed work before definite conclusions can be reached. Solomon has already shown that little evidence exists of long pluvial periods in East Africa, yet, until now, these have been the basis and framework of the East African Pleistocene succession, and have led, not only to local correlations, but also to those of far wider significance, involving ice-ages and great astronomical phenomena. Similarly, the simple fact that some cultures covered large areas for a considerable length of time, has led to the postulation of very close correlations between the European and African sequences. Such correlations, while apparently simplifying the chronology, have actually resulted in confusion and erroneous conclusions concerning race, migrations, land-bridges and the like, in spite of conflicting evidence sometimes supplied by fauna and local geology. In a word, it is scientifically impossible, at present, to effect more than the broadest correlations between Stone Age cultures over large areas, and, for this purpose, it is absolutely necessary to distinguish between long-lived cultures of great extent and local industries of, probably, only tribal significance.

CHAPTER V

Outline of the Uganda Culture Sequence

IN this short description of the Uganda culture sequence, I have kept to chronological order, although it makes some repetition inevitable, because of the persistence and overlapping of some of the cultures. In later chapters, each culture, with its subdivisions, will be dealt with as a complete entity, accompanied by the geological evidence necessary for its dating. Thus, all the information relating to one particular culture will be kept together.

All descriptions of tools, geology, climate, etc., are based on our own collections and observations in the field and afterwards, unless otherwise stated.

A few remarks on dating are necessary before embarking on the brief description of the chronological sequence.

The earliest Pleistocene deposit in Uganda, fixed by Palaeontology, is the Kaiso Bone-Bed, which Dr Hopwood has dated to the Lower Pleistocene. This dating, following Haug's definition, depends on the presence of true elephants, horses and oxen and, according to it, deposits such as the Cromer Forest-Bed become Middle and not Lower Pleistocene in age. Similarly, the beginning of the Oldoway succession must belong to the Middle Pleistocene.

As the Uganda culture sequence begins only after some unknown lapse of time after the Kaiso Bone-Bed, I have been obliged to regard the earliest cultures as Early-Middle or Middle Pleistocene in age, although, on the more usual method of dating, they would be considered as Lower Pleistocene.

THE LOWER KAFUAN PEBBLE-CULTURE

The Kafuan makes its first appearance very heavily rolled in the Peneplain Boulder-Bed belonging to the end of the Epi-Kaiso, or third series of lacustrine deposits in the Albertine Rift, and in the 50 ft. $\pm$ terrace of the Muzizi river, above the Rift. As far as we know, this is the oldest human culture in East Africa, being long pre-Chellean and older even than the Oldowan in origin. It is dated to the Early-Middle Pleistocene.

The Lower Uganda Oldowan

This stage of the Uganda Oldowan culture was also found in the Epi-Kaiso Peneplain Boulder-Bed and in the Muzizi 50 ft. terrace gravels which contain Lower Kafuan. The Oldowan is quite definitely younger than the latter and is entirely different in technique and type. It might have been permissible to argue that the Lower Oldowan was a more developed stage of the Lower Kafuan were it not for the fact that we have much later Kafuan and Oldowan industries, both of which preserve their cultural characters, although they were contemporary, showing that they were always separate entities.

In both the Epi-Kaiso Peneplain and the Muzizi 50 ft. gravels, the Lower Oldowan tools are much less heavily rolled than the Lower Kafuan and must, therefore, be later. For the same reason, it is probable that the Lower Kafuan is older than the wet episode responsible for the Epi-Kaiso and other similar gravels, and had already suffered some wear and rolling before being included in these.

The Early Chellean

Proof that the Early Chellean was later than the Lower Oldowan was also provided by the deposits in the Albert Rift, for Early Chellean tools are to be found heavily weathered, but not water-rolled, among the Epi-Kaiso Peneplain gravels, made on boulders from this bed. In their simplicity of technique and extremely primitive appearance, these hand-axes are comparable with those in the Early Chellean of Oldoway and, perhaps more closely, the oldest Stellenbosch of South Africa.

Elsewhere in Uganda, the same stage is found in the Older Rubble—formed during a very dry climate—which underlies the Younger Rubble, containing Lower Acheulean.

At present, we know of no other later stages of the Chellean culture, and such evidence as there is suggests that Uganda was not inhabited by Chellean man after his initial visit. This may have been due to unfavourable climate.

56

Outline of the Uganda Culture Sequence

THE LOWER ACHEULEAN

The Chellean Older Rubble is overlain by another similar deposit which began to form after an interval indicated by a break in the accumulation of angular rock fragments. In favourable spots this break is represented by clayey soils which are interpreted as being the result of a moist climate, while the rubbles are essentially "arid" deposits.

In this Younger Rubble, whose formation continued, with possible short interruptions, for a considerable length of time, there is a variety of Stone Age material, comprising several different cultural and industrial stages. The only satisfactory method of estimating their relative ages was the study of their physical condition. By this means, we could check the results of division by typology.

The oldest tools in the Younger Rubble are some very weathered hand-axes and flakes which I have classed as Lower Acheulean; my reasons will be given more fully in the appropriate chapter, but I may say here that both typology and state of preservation support this conclusion.

THE MIDDLE ACHEULEAN, STAGE A

There are two parts to the Middle Uganda Acheulean, an earlier, Stage A, and a later, Stage B.

Stage A belongs to the time when the Kagera valley was flooded for a considerable distance by reason of a slow, westerly land-tilt away from Lake Victoria. The valley thus became a long, narrow arm of the lake. High-water level is represented by a bench cut in the Older Rubble and in part of the Younger and, in some few places, by well-sorted beach-gravels. At Nsongezi, we have a few implements from the hardened beach-gravels, among which are a number of Oldowan type. Close by, an abundance of similar tools formed a layer on the floor of the "lake", overlying sediments which now constitute the lower part of the 100 ft. terrace series. Owing to its position in the floor of the valley, this bed was at first thought to be the same as the M-Horizon, though the latter contained much better-made implements which are fresh, while the former is a gravel in which all the tools are rolled and of cruder facies. It would seem that the Stage A people lived just above the

57

level of the beach and that some of their artifacts were washed down into the valley bottom, where they formed a layer.

THE UPPER OLDOWAN

As I have just said, the beach and valley gravels contain Oldowan as well as Acheulean types. In fact, these typical core-choppers constitute 83% of the total. Owing to the similar state of rolling in the Oldowan and Acheulean implements, there appears to be no reason for supposing the former to be older, and it seems likely that they belong to a late stage of the Oldowan culture which came into contact with the Early-Middle Acheulean and was ultimately absorbed by it. In the developed Middle Acheulean of the true M-Horizon, the Oldowan tools still persist, though somewhat modified in the technique of manufacture.

THE MIDDLE KAFUAN

At another site in the Kagera valley, several miles from Nsongezi, the Early-Middle Acheulean beach contains nothing but an abundance of small pebble tools of Kafuan type.

This beach was clearly the result of a south-westerly tilt towards the Albertine Rift Valley, which must also have reversed others of the easterly-flowing rivers. The Kafu is an example of this, for, having flowed towards the north-east ever since the original (Pliocene?) reversal, proved by Wayland, a direction of flow back towards the Rift is shown by the gravels of the 50 ft. ± terrace, that is, there must have been a south-westerly tilt some time before these were laid.

This terrace contains a series of tools very similar to those found in the Kagera beach and this fact, with the evidence of south-westerly tilting, provides fairly conclusive proof of their contemporaneity. We know of no other south-west tilt, all other movements having been towards the north-east.

Wayland postulates pluviation, instead of tilting, to account for this river reversal, that is, the filling up of the lakes, due to increased rainfall and their encroachment up the major valleys until these overflowed at the divides. There are several reasons against this, chief among which is the lack of evidence of heavy rainfall in places where it should have caused the formation of boulder-beds and so on. Also, the level of the

Early-Middle Acheulean beach is quite out of keeping with a valley invaded merely by a flooding lake.

Thus, both physiography and archaeology suggest a similar age for the Kagera valley beach industries, Early-Middle Acheulean and Middle Kafuan, and the Kafu 50 ft. terrace stage of the Kafuan.

THE UPPER ACHEULEAN

There is only one industry in the M-Horizon Rubble in the Kagera valley, and we may suspect that after a time the increasing aridity drove Acheulean man towards more favoured spots. Lake Victoria was the obvious place to make for, as it is unlikely that it ever dried up altogether during this period. Unfortunately, we know of no exposures of the old beaches or shores of this time, in which there might be tools, because such areas were probably subject to the widespread flooding consequent on the last, north-easterly tilt, from which parts of the lake have never recovered. Thus, the immediate, post-M-Horizon shores and beaches will have been covered by later deposits.

There is, however, a rubble near Jinja (see p. 67, fig. 3) which contains, among other industries, some rather advanced Acheulean hand-axes whose state of preservation is distinctly earlier than most of the rest, which are mainly Levalloisian. The nearest counterpart of this hand-axe group is the Upper Acheulean of Kenya and Oldoway, of Stage 5 type.

Apart from this site near Jinja and one other, in the Sango Hills, I know of no place in Uganda where the Upper Acheulean occurs, though it may be present in other favoured places near the lakes. It is entirely absent in the Kagera valley, for instance, except on the Sango Hills, near the mouth of the river. Here, again, the small, Upper Acheulean hand-axes were found in a rubble and are clearly older than the large, crudely made "Sangoan", that is, Proto-Tumbian, industry that accompanies them.

We may regard the Middle Pleistocene period as ending with the Upper Acheulean.

THE LOWER LEVALLOISIAN

Previous work in East Africa has indicated that there was some considerable break in both the geological and cultural sequence between

Kamasian and Gamblian times.[1] For reasons which will be given later, I feel that the magnitude of this break may have been exaggerated, and further work may show, for instance, that the *Final* (not Upper) Acheulean actually bridges the gap between the two periods in some places.

However that may be, there is no sign, in Uganda, of Final Acheulean in the riverine or lacustrine deposits of post-Upper Acheulean age, and the new phase immediately produces Levalloisian industries. From now on, this culture continues to exist for a considerable period and is to be found in almost every corner of the country.

The Lower Uganda Levalloisian makes its appearance in riverine sands overlying the M-Horizon in the Kagera valley. Clearly, the rainfall was now normal again and the rivers flowing once more. There are three horizons in these sands, and a statistical study of the tools shows a gradually increasing proportion of distinctive Levalloisian characters.

UGANDA PROTO-TUMBIAN

The deposition in the Kagera valley of sands with Levalloisian was terminated by a slight oscillation to drier conditions and the creation of a land surface—the N-Horizon. Levalloisian tools are also present here, but the horizon is chiefly remarkable for the appearance of an early stage of a new culture, the Tumbian.

Large, clumsy hand-axes and picks, cores and lumpy flakes, with but little retouch, represent the beginnings of a culture whose beauty, at a later stage, was never surpassed at any period in the Stone Age history of Uganda.

MIDDLE LEVALLOISIAN I

The N-Horizon stone-bed, or rubble, lies at the base of a bed of thin alternating layers of fine white sand and clay. It is of uneven thickness and probably represents a time of sporadic and irregular rainfall. In the upper part of this deposit, which we call collectively the N-Horizon Bed, there is no sign of Proto-Tumbian tools, but we found two more thin horizons of Levalloisian. As the N-Horizon offers a convenient

---

[1] These terms are used only in a geological sense, to indicate the age of the deposits, not climatological phenomena.

change of sedimentation for subdividing the geological succession in the upper part of the Kagera 100 ft. terrace, and the Levalloisian tools are also further developed than those from the sands below, I have called these N-Bed stages Middle Levalloisian I.

MIDDLE UGANDA TUMBIAN

Although there is no apparent break in the Levalloisian sequence throughout the upper part of the Kagera 100 ft. terrace, there is no sign of the development of the Proto-Tumbian in the sands and clays of the N-Bed, and its presence at the base of these seems to indicate an incursion into the area, probably of migratory significance.

In the uppermost clays, above the N-Bed, however, there is an abundance of developed Tumbian artifacts. Hand-axes and picks are still numerous, but much better made than those of the N-Horizon, while a variety of beautiful, leaf-shaped tools and tranchets show that the culture was considerably further advanced than in the Proto-Tumbian stage. There seems to be no doubt, therefore, that this second stage, which I call Middle Tumbian, represents a new wave of this culture, which had had time to develop since the earlier stage in some place outside this area.

MIDDLE LEVALLOISIAN II

The Tumbian clays of the Kagera 100 ft. terrace also contain large numbers of well-made Levalloisian tortoise-cores and prepared flakes. These show so little advance on those from the N-Bed that I have called them Middle Levalloisian II, rather than Upper Levalloisian.

There is no trace of any mutual borrowing between the Tumbian and Levalloisian industries, despite their absolute contemporaneity and close proximity in the valley. Such Tumbian tools as are made on flakes never have prepared striking platforms, nor do Levallois flakes with such platforms ever exhibit the flat, wood-flaked retouch characteristic of the Tumbian. I do not deny the possibility of mixture, but we have no evidence of it so far.

LATE KAFUAN

The latest pure Kafuan is that found so abundantly in the Kafu "flats" terrace and in open stations above it, in brickearth. It also occurs near

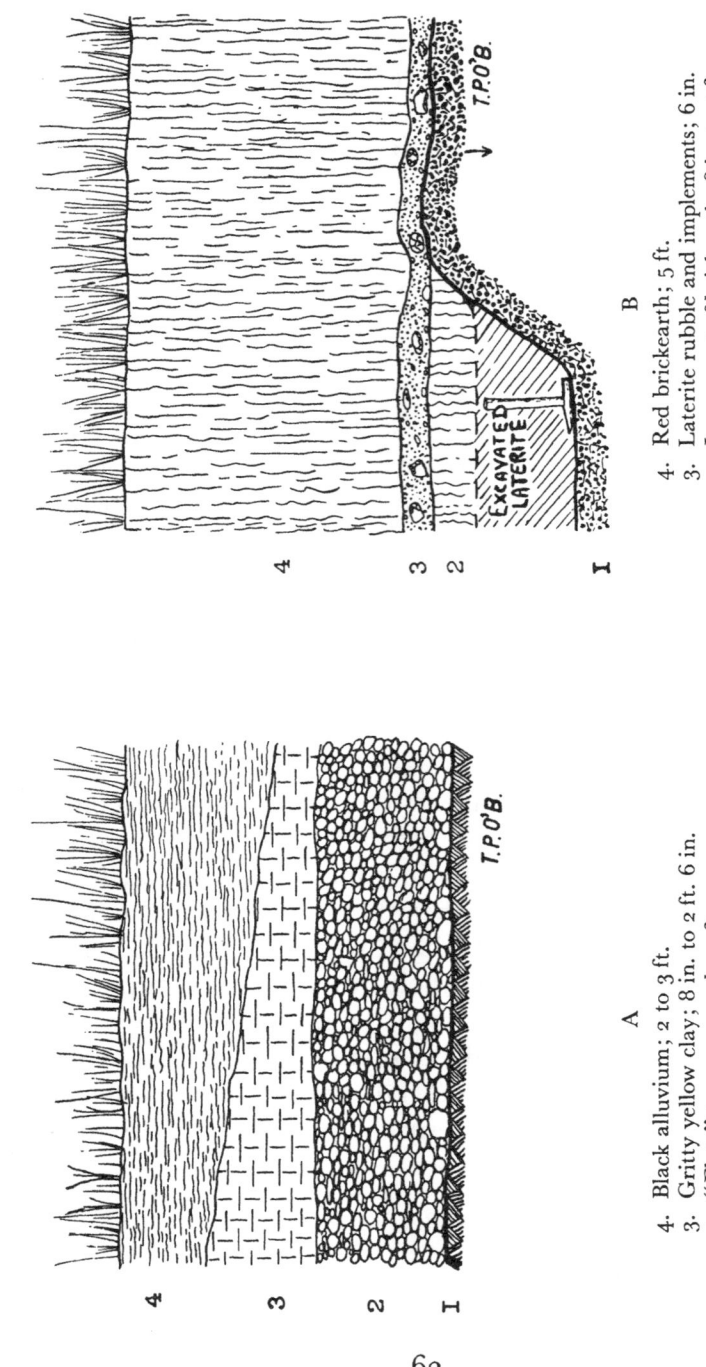

A

4. Black alluvium; 2 to 3 ft.
3. Gritty yellow clay; 8 in. to 2 ft. 6 in.
2. "Flats" terrace gravels; 3 ft.
1. Rotted bedrock.

B

4. Red brickearth; 5 ft.
3. Laterite rubble and implements; 6 in.
2. Lower part of brickearth; 8 in. to 2 ft.
1. Solid laterite; down to bedrock.

FIG. 5. A. Section on "flats" terrace, Kinoga, Kafu River.
B. Section at Camp Site, above "flats", Kinoga.

62

Jinja, at the north end of Lake Victoria, in sediments which were probably laid down during the flooding caused by the last north-easterly tilt and just before the Ripon outlet was established.

This period is fairly clearly dated because, in the Kagera valley, the tilt was directly responsible for the permanent reversal of the river back to Lake Victoria and the cutting of a new channel through the 100 ft. terrace deposits. Thus, it ended the silting phase at the top of this series which contains Middle Tumbian and Middle Levalloisian II.

At Jinja, Late Kafuan tools occur down to 2 ft. in a hard, ironstone deposit, whose water-laid origin is demonstrated by the bedding planes and pebble zones in it. On the surface, also firmly embedded, we found a number of typical Levalloisian cores and flakes, while, in a detrital deposit of lateritic rubble surrounding the solid laterite, there were many more tools, including the two former industries and Upper Tumbian. Thus, there seems little doubt as to the approximate age of the Late Kafuan here.

The same north-easterly tilt also reversed the river in the Kafu valley, first causing erosion and then the laying of the "flats" terrace gravels. The implements from this terrace are very similar to those at Jinja, and I have no doubt that they belong to the same stage, Late Kafuan.

Other sites in red earth (brickearth) above the "flats" produced exactly similar artifacts, in mint-fresh condition. If any further proof were needed, beyond that provided at Jinja, of the comparative youth of this stage, it would be given by this discovery of typical, "flats" Kafuan fresh in brickearth, which, in Uganda, is invariably a young deposit.

UPPER LEVALLOISIAN

The old shore-deposit, near Jinja, containing Late Kafuan tools, gradually became hardened, but, before lateritisation was complete, Upper Levalloisian people came into the area and some of their implements are to be found firmly embedded in the topmost part of the deposit. Similar tools, both with and without lateritic adhesions, occur in the minor, detrital rubble round about.

UPPER UGANDA TUMBIAN

Tumbian implements of later type than those of the Kagera 100 ft. terrace clays were also found in the rubble described in the foregoing paragraph. Some of these bear lateritic adhesions and probably belong to the same age and deposit as the Upper Levalloisian with similar incrustations, that is, the final stage of lateritisation of the old shore deposits, containing Late Kafuan.

Upper Tumbian and Upper Levalloisian tools that post-date the period of lateritisation are also plentiful in this rubble, and it is possible to divide these fairly satisfactorily on the basis of their relative states of preservation, particularly by staining. Any object in contact with this iron-charged material for any length of time, acquires a tinge, pale or dark, according to the time involved.

STILL BAY: WALASI VARIATION

Up to now, it has been generally accepted that the Still Bay "culture" was the outcome of contact between Late Levalloisian and a "Neoanthropic", that is, blade and burin culture.[1] There is no need, here, to go into the reasons why I dissent from this view, except to say that we have a form of Still Bay in Uganda, which shows no sign of this particular culture contact. On the contrary, the industry seems to develop directly out of the Upper Levalloisian, but there is a strong possibility that the *biface* tools that characterise it were the result of the *borrowing* of a *biface* technique from the Tumbian or some related culture.

This form of the Still Bay is well represented in the neighbourhood of Walasi Hill, west of Mount Elgon, Eastern Province, and near Jinja. In both areas the tools occur in a rubble, not in a stratified deposit, but, as I have already remarked, it is possible to get some idea of the relative ages of the implements from their state of preservation and staining. At Jinja this method showed that the Early Still Bay followed the pure Upper Levalloisian and Upper Tumbian, while, at Walasi, though the only industry was Early Still Bay, the same method enabled us to see reasonably clearly how it developed.

MAGOSIAN

This culture[2] offers an interesting example of a contact between the *Late* Still Bay and some blade and burin industry. The apparent evolu-

---

[1] Leakey, *Stone Age Africa*, 1936, pp. 62 *et seq.*; Burkitt and Wayland, The Magosian Culture of Uganda, *J.R.A.I.*, vol. LXII, July–December, 1932, p. 379, and several others.
[2] The Magosian Culture of Uganda, *J.R.A.I.*, vol. LXII, July–December, 1932.

tion at the type-station of Magosi, from rather crude beginnings to more developed microlithic tools, suggests that there is an early phase of the culture there, and this is supported by finds of more advanced Magosian in Abyssinia,[1] and Tanganyika.[2]

The culture of the type-station has been dated to the dry phase following the Gamblian in Kenya, but the Late Magosian is probably Makalian, or later, in age. It is considered to be the parent of the East African Wilton B.

We ourselves found a curious little industry of microlithic type, in a shelter in West Uganda, including degenerate Still Bay points but few true Wilton forms, such as the Magosian possessed from the first.

LATE UGANDA TUMBIAN

This stage is later than the Upper Tumbian and occurs sporadically in Uganda, chiefly in the west. It is small, almost microlithic, and usually made of white quartz; it appears to be quite late. We have no means, at present, of dating it more precisely than by saying that it is probably as late as the peculiar Still Bay derivative of Magosian type, noted above.

KAGERAN

This is an industry which appears to have no counterpart in East Africa. It bears some resemblance to some of the Strandlooper industries of South Africa, also of late date, but no connection is suggested, hence the necessity for a local, distinguishing name.

The industry was found on the 30 ft. terrace of the Kagera, and a few tools of the same type (chipped quartzite pebbles and rough flakes) had also been found by Wayland, at the base of the Wilton-Neolithic layers in a rock-shelter nearby.

WILTON-NEOLITHIC

This is a microlithic culture which is best represented in cave and rock-shelter sites such as the shelter at Nsongezi, discovered and first tested by Wayland. Another, somewhat similar industry was found by us in a

---

[1] Breuil and Kelley, *Journ. de la Soc. des Africanistes*, Tome VI, 1936, pp. 111–40.
[2] Leakey, *Stone Age Africa*, 1936.

PLATE IV

Fig. 1: top left. Early-Middle Acheulean gravel bed at a borrow pit at Mile 51 on the Kampala-Jinja Road. It is the equivalent in age of the Phase A gravels in the M-Horizon at Nsongezi. The gravel lies at the base of a thick bed of laterite and overlies rotted quartz bedrock.

Fig. 2: top right. Surface of the Younger Rubble below stoneless brickearth near Bugungu, Lake Victoria. The rubble produced very heavily weathered and rotted Middle Acheulean tools, of the same type as in the Phase B part of the M-Horizon at Nsongezi.

Fig. 3: bottom left. A rubble horizon at the Railway borrow pit, near Bugungu, which produced Upper Acheulean, Proto-Tumbian and Levalloisian tools. It is overlain by brickearth containing Neolithic remains.

Fig. 4: bottom right. A view down the Napoleon Gulf to the Ripon Falls and Nile, at Jinja.

shelter at Walasi Hill, near Mbale. While the culture contains a number of thumb-nail scrapers of Wilton type, the other tools are not typical East African Wilton forms. Perhaps the best way of describing the complex is to say that it contains elements of both the Wilton and Gumban of Kenya. Until, however, we know more about the latter as well as this Uganda material I prefer to call the latter Wilton-Neolithic.

CHAPTER VI

The Kafuan Pebble Culture

THIS culture was discovered in 1919, by Wayland, in the terrace gravels of the Kafu River. Some later finds in the same locality and the connected geology were described in the *Annual Reports* of the Geological Survey for 1926 and 1927. Further reference to this culture was made by Wayland in the *Transactions of the Royal Society of South Africa*, vol. XVII, part 4, 1929, and in "Rifts, Rivers, Rains and Early Man in Uganda", *J.R.A.I.* vol. LXIV, July–December, 1934.

GEOLOGICAL DATING

The work of the Geological Survey of Uganda had proved that the Kafu River possessed five terraces, at the following heights above the present river:[1]

225 ft. ±, some 20 ft. below the 245 ft. peneplain;

175 ft. ±, the equivalent of the Singo Plain;

50 ft. ±; "flats", just above the present swamp; "subflats", below swamp-level and determined only by boring.

Stone tools were reported from all of these, except the last.

Originally, these terraces were considered to "*represent the dissection of a peneplain which was in existence at the end of Pliocene times*",[2] and to cover the whole of the Pleistocene period down to recent times. Later, however, Wayland[3] correlated them with his Pluvial I, thus making both the terraces and their contained artifacts entirely pre-Chellean, that is, Lower Pleistocene in age, or, on our own dating, Early-Middle Pleistocene.

The levelling of the terraces carried out by the Survey seemed to prove Wayland's original hypothesis of river-reversal by earth movements connected with the Rift Valley, and these results are described in the 1927 *Report* as follows:

...*it is found that these* (the terrace gravels) *reveal two remarkable buckling*

---

[1] As some of the terraces dip and are crossed by more recent ones, the heights given are, necessarily, only the highest points originally levelled.

[2] *A.R.G.S.U.* 1927, p. 18.   [3] *R.R.R.E.M.U.* table.

movements to the north in Pleistocene times. So far as it is possible to show from the area studied, the 245 ft. peneplain appears to parallel the 225 ft. terrace. That movements were initiated in Post-225 ft. times is proved, however, by the fact that although for about 25 miles this terrace is parallel with the 175 ft., beyond that point the two terraces diverge due to the greater dip of the 225 ft. It can be shown that this first great buckling movement probably caused a reversal of drainage—that is, from an original south-westerly direction into Lake Albert, to the present north-easterly course into a new lake (the remnants of which form the present Kioga) brought into existence by the movement. It is probable that there was continuous slight movement through 175 ft. times, but there was undoubtedly a second great buckle after the deposition of the 50 ft. gravels. The formation of the 225 ft. and 50 ft. terraces is ascribed to rejuvenescence due to pluvial periods in the tropics (corresponding with glacial periods in higher latitudes) while the 175 ft. terrace and present flat has been brought into being by tectonic movement.[1]

This information has been given here because it includes almost the only evidence of the age of the Kafu terraces and their contained artifacts. The movements described above were, of course, of more than local importance and, as I have said in an earlier chapter, led to the reversal of other rivers, besides the Kafu, as Wayland had already realised.

The tools themselves provide no evidence of their age as compared with other Palaeolithic cultures. They are merely rather small, simple chopping and scraping implements, whose size was probably regulated, at least in part, by that of the available raw material. It is also most unfortunate that there is no other Lower Palaeolithic culture in this area which could serve as a basis for age correlation, so that, apart from the geological implications of the tilted terraces, there is nothing whatever to guide us. If the movements that reversed the river or led to its rejuvenescence were very early events, then the tools must also be early. If, on the other hand, the movements can be proved to have occurred within later Palaeolithic times, the Kafuan must be of this date. In this connection, the Kagera valley proved illuminating, for it provided evidence of several land tilts and consequent river reversals that are

---

[1] Solomon, however, offers an alternative explanation for the formation of the low divides in such two-way rivers (see p. 23).

fairly well dated by abundant Stone Age cultures, among which the Kafuan is represented.

First of all, the type area must be considered. It has been shown that the first movement probably reversed the Kafu from its original south-westerly to a north-easterly direction of flow, after the deposition of the 225 ft. terrace gravels, and that there was a second movement in the same direction after the 50 ft. terrace. But the Survey's original work showed that the 50 ft. gravels themselves contained internal evidence of a direction of flow towards Lake Albert,[1] showing that an intermediate reversal back towards the south-west had taken place since the 175 ft. terrace. Hirst remarks:[2]

> ...the increase in thickness of the 50 ft. gravel in a down-stream direction is to be noted. It is 3 ft. at the Kafu Bridge, 6 ft. 6 in. at the 2-mile line and 5 ft. 6 in., at the 4-mile line. At Kinoga it has increased to 16 ft. of very coarse (3 in. to 4 in.) pebbles without appreciable binding, and is still 16 ft. at Lusensa. Two significant facts stand out therefore—first, a decrease in height of the 50 ft. terrace, and second a coarsening and thickening of the gravel in an opposite direction to that in the gravel of the Kafu flat. It is clear, therefore, that between the deposition of the gravels of the 50 ft. terrace and the Kafu Flat, there was a decided tilting of the land to north-east—a tilting which must have resulted in the reversal of the Kafu River.

It is strange that this fact (the south-westerly flow of the 50 ft. river) is not spoken of again in the 1927 *Report*, but only the first (post-225 ft.) and last (post-50 ft.) reversals are mentioned. We verified the facts of the thickening and coarsening of the 50 ft. gravels in the present down-stream direction.

Hirst reported[3] that all the terraces (except the subflats) yielded artifacts, but he did not say whether or not the tools were rolled and, therefore, at least as old as the gravels, though he implies that they were as old as their respective terraces. As we did not examine the 225 ft. and 175 ft. terraces ourselves, I do not propose to say anything about the industries reported from them, except what may be inferred from other evidence.

---

[1] Solomon, however, considers the 50 ft. gravels to be more in the nature of a beach (see p. 38).

[2] *A.R.G.S.U.* 1926, p. 15.   [3] *A.R.G.S.U.* 1927, p. 18.

This evidence is purely circumstantial, at present, and is based on the fact that the tilts were of regional magnitude. We know that the earliest rift-genetic movements were long anterior to the Lower Palaeolithic, because a whole series of non-implementiferous deposits were laid within the Rift, and these *end* with a bed containing the earliest known pre-Chellean artifacts. It is now almost beyond question that the gradual genesis of the Rift Depression was accompanied by slow uplift along its margins and that the uplift expressed itself as a tilt eastwards, away from the depressed zone. This suggests an early date for the first river reversal, and makes it likely that the 225 ft. westerly flowing Kafu was earlier than the first major movements of rift genesis. Judging by the thickness of the pre-Chellean Albertine deposits, such westerly flowing rivers must be of Pliocene age, for, by the time of the Lower Pleistocene[1] Kaiso Bone-Bed, the rift had long been in existence, much as it is to-day, while stone tools do not appear until considerably later than Kaiso Bone-Bed times.

Thus, it seems unlikely, for several reasons, that Man was in existence at the time of the Pliocene, 225 ft., westerly flowing river.

The 175 ft. terrace may be of any age subsequent to the post-225 ft. tilt reversal and prior to the pre-50 ft. *return* tilt reversal, but there is no evidence for more precise dating than this at present. The age of the pre-50 ft. tilt, however, is reasonably well attested in another part of the country, where it led to the ponding up of a valley connected with Lake Victoria, and its probable eventual reversal towards the Albertine Rift. Thus, from being tributaries of the present Kioga and Victoria drainage basins, these rivers became their outlets, of greater or lesser significance according to the degree of gradient reversal. In the Kagera valley, this tilt was post-Chellean, and in the highest beach produced by the flooding of the valley by Lake Victoria at that time, pebble tools of Kafu 50 ft. terrace type have been found both by Wayland and ourselves. At other sites in the valley, the same beach produced rolled Early-Middle Acheulean implements.

LOWER KAFUAN

The oldest datable Kafuan stage is that occurring in the 50 ft. $\pm$ terrace of the Muzizi River and in the equivalent Epi-Kaiso peneplain gravels

[1] Upper Pliocene according to the older method of dating the fauna.

in the Albert Rift. The 50 ft. terrace of the Muzizi is the only one recognisable in that valley, though isolated pebbles occur at various heights up to the old peneplain level, some 800 ft. or more above the river. The terrace was cut out of the soft, weathered, upper part of the Muzizi sandstone.[1] The gravel is large and unevenly sorted, and has every appearance of having been laid by a torrential river, fed by considerable rainfall (see p. 30).

The tools in this gravel were first noted by Wayland in 1919; he called the whole series Developed Kafuan, and recently equated it with the Oldowan of Leakey.[2] Actually, there are two cultures present, of which the younger is more advanced and more abundantly represented than the older. As far as the younger series is concerned, Wayland was right, it is Oldowan, but the older group is both more heavily rolled and of more primitive appearance, and is clearly of Kafuan type.

By themselves, both these groups would have been almost useless for dating purposes, just as in the case of the industries at the Kafu, but, whereas the latter can only be dated tentatively, with reference to tectonics, the Muzizi series can be placed in the sequence definitely, by archaeology, from data obtained elsewhere. This consists of the succession that we established in the Toro-Semliki area, the south-east corner of the Albert Rift Valley. It is not necessary to repeat in detail the geological succession given by Solomon in Chapter III, but simply to point out that great thicknesses of deposits are preserved in this area, covering the whole history of the lake, from its inception within the post-Miocene trough to the just pre-Chellean, Epi-Kaiso lake. The latter was responsible for the deposition of the Epi-Kaiso series, which were laid against the faulted Kaiso Beds. At the top of the Epi-Kaiso series is a boulder-bed, of such widespread character that we call it the Peneplain Boulder-Bed.

It was evidently laid on the floor of the final Epi-Kaiso lake, after torrential streams had brought the material down from the high ground round the basin. In almost every particular, this boulder-bed is similar

[1] Probably the equivalent of the Albertine Kisegi series, laid in the floor of the valley when it was graded to the pre-major-fault Albert trough, probably in Lower Pliocene times.

[2] *R.R.R.E.M.U.* p. 336.

to the Muzizi 50 ft. terrace, especially in its tool content, which consists of two series, older and younger—Kafuan and Oldowan.

Ultimately, the lake floor became a land surface, and the boulders an inexhaustible source of raw material for all later peoples. The first folk to make use of it were the Early Chelleans, whose large, coarse, simply made hand-axes, heavily weathered, but never water-rolled, can be collected in almost every place where the boulder-bed is exposed. Thus, the pre-Chellean age of the Epi-Kaiso Peneplain Boulder-Bed and the Muzizi 50 ft. terrace, and their two contained cultures is established.

Typology

The Lower Kafuan tools are very heavily rolled and almost invariably made on fairly large (4–5 in.) pebbles of quartz. The majority are oval and comparatively flat. They include crude cutting and chopping implements, very simply made by the removal of a few flakes from one end or side. The extent of the rolling makes it difficult to see whether the tools had ever been more finely retouched after the primary flaking, but, judging by later Kafuan industries, they were not. The working is, however, certainly human, as the Abbé Breuil and others have agreed.

In addition to the flattish, oval implements, there were a few of the spherical type usually associated with the Oldowan, but this indicates, not that the Lower Kafuan is an early stage of the Lower Oldowan, or has anything to do with it, but simply the fortuitous use of spheroidal pebbles and boulders. In the same way, in the Oldowan of the type station, "Kafuan" types occur, obviously because of the chance selection of flattish pebbles. In view of the persistence of the Kafuan of characteristic type into much later (Acheulean) times, when it was a contemporary of a later stage of the Oldowan, it seems most unlikely that the older Muzizi River industry is merely a primitive form of the Oldowan. At this, as at all other Kafuan sites, the characteristic Kafuan technique resulted in tools that are broad and long in comparison with their thickness, whereas the Oldowan type tool is essentially a spherical or cuboid form. In our collection, the proportion of Oldowan shapes in the most heavily rolled group was only 26%; the rest were all ordinary Kafuan types. Thus, I call this industry Lower Kafuan, although there is a bare possibility that there may be an earlier stage at the Kafu River.

MIDDLE KAFUAN

The industry that I propose to call Middle Kafuan is that of the 50 ft. ± terrace of the Kafu River and the so-called 220 ft. ± terrace of the Kagera. Dealing with the latter first, it is necessary to point out that this terrace is really a beach-gravel which was laid when Lake Victoria flooded up the Kagera valley after a tilt towards the south-west.

Further detail will be found in the chapter on the Middle Acheulean as to the reasons for this beach being dated approximately to that time. Its position, far above the 100 ft. ± terrace, at first led us to believe that it was pre-Acheulean, though post-Early Chellean in age. Later on, however, when Early-Middle Acheulean implements were found on it, near Nsongezi, we realised that it could not be a river terrace, but must be a lake-beach, whose height, so far above the 100 ft. deposits, was due to deep drowning by Lake Victoria, as a result of a land tilt.

The most imposing remnant of this beach occurs in Tanganyika, just over the border, about 15 miles downstream of Nsongezi.[1] By a lucky chance, the beach-gravels here were deposited on an outcrop of rock that had resisted erosion during the earlier history of the river. This outcrop now forms a considerable island, detached from the main wall of the valley and rising steeply out of the 100 ft. terrace plain. At the riverward edge it is capped by a gravel whose date is impossible to establish, as it contains no tools. It is the only remnant we know of the top, 270 ft. ± Kagera terrace (see p. 36, fig. 4).

Some 50 ft. lower than this, where the island slopes back to the hills, lie the 220 ft. beach-gravels. The main road cuts across them and provides the small, 4 ft. section shown on p. 47, fig. 2. The gravel is small and extremely well-sorted; it contains numerous quartz pebble tools of Kafu 50 ft. terrace type, all heavily rolled. There are also a number of quartzite flakes, which I was, at first, at a loss to account for in such an industry. When we realised the comparatively late date of the beach, however, it was obvious that this deposit could contain small tools, flakes and general debris from whatever sites it had cut across, and

[1] At Mile 24¾ on the Nsongezi-Kagera Port Road.

that the small, quartzite flakes probably belonged to one of the two other cultures of the same period. Had the beach-gravels been larger at this point, I have no doubt that we should have found the larger tools as well, but water-sorting had resulted in only the smaller Kafuan pebble tools and other debris being included in it. At this point, it must be remembered, the beach lies at least a mile away from the hills rimming the valley. At Nsongezi, however, a small remnant of the same beach[1] lies at the foot of the hills (see p. 49, fig. 3), and this contains larger tools belonging to Upper Oldowan and Early-Middle Acheulean industries. Kafuan specimens have not so far been found at this point, perhaps because the makers of this culture did not occupy this part of the valley.

Kafu 50 ft. ± Terrace

Earlier in this chapter, reference was made to the direction of flow of the Kafu 50 ft. river, as indicated by the thickening and coarsening of the gravel in a present downstream direction. As the previous, 175 ft. ± river apparently flowed in the opposite direction, there must have been reversal between them. This was the result of a tilt back towards the Albertine Rift, that is, the relative lowering of the uplifted area bordering it, perhaps due to slipping or settling of the uplifted zone along old fault-planes.

This is the only instance we know, in this valley, of a return tilt back towards the Rift, and only one such return is recognisable in the Kagera valley, where it caused flooding by Lake Victoria and the formation of the 220 ft. ± beach. This evidence, and the similarity of the implements from the two deposits, make their approximate contemporaneity almost certain.

The tools obtained from this terrace are rather few and not at all advanced. They are, in fact, very little better than those of the Lower Kafuan. Like those from the Kagera, they are mainly small, though quite sizable pebbles are common in the terrace; so the dimensions of the artifacts are evidently not entirely dependent upon those of the

---

[1] Solomon is inclined to think that there may be several different beaches in the valley, corresponding to various rest-levels of the lake during its pre-M-Horizon occupation of the valley, but only a great deal of levelling could decide this point.

available material but may be of morphological significance, that is, a definite preference for tools of a certain type and size, which varied at different periods in the history of the culture. It is quite possible, however, that if home-sites of this stage are found, tools of all sizes will occur.

Typology

There is little to be said about the individual tools from the Kafu 50 ft. terrace and the Kagera 220 ft. beach. They are best described as chipped pebbles which suggest small, chopping, cutting or scraping implements of no specialised forms. It would be valueless to attempt to subdivide them into tool classes, for every gradation from one "class" to another can be seen, from obvious scrapers or choppers to "points", "chisels" and so on.

LATE KAFUAN

Ever since the Kafu 50 ft. terrace stage, the Kafuan seems to have gone on developing in some parts of Uganda, and to have bridged the supposedly long gap between the Lower and Upper Palaeolithic of East Africa. By the time it appears once more, cultures such as the Levalloisian and Tumbian had already established themselves in the country, but, as in much earlier times, the Late Kafuan shows no sign of extraneous influences. It remains essentially a simple pebble culture that found no use for "biface" or flake techniques. Such exclusiveness for so great a length of time is suggestive of some strong, racial, perhaps even generic difference between the Kafuan people and all others, which made any but the most temporary contact impossible. There is, however, some rather slight evidence that, towards the end, this culture became influenced by the Levalloisian, and may even have been submerged by it. However that may be, the Kafuan is still the dominant culture in the Kafu valley in Upper Levalloisian times, and, at Jinja, 135 miles away, it actually kept out the Levalloisian culture for a time.

It will be remembered that the Kafu 50 ft. terrace gravels show a south-westerly direction of the river's flow. The "flats" terrace, however, shows that the river was again flowing north-east, towards Lake Kioga, and the present, easterly dip of the 50 ft. terrace, until it is

77

crossed and cut out by the "flats", indicates that the reversal was due to a north-easterly tilt, away from Lake Albert. All the evidence shows that this was the last tilt to effect important changes in the courses of the Uganda rivers, and their present valleys and directions of flow are due to it.

The same tilt caused the rapid draining away of the water from the ponded-up Kagera, and its consequent rejuvenation. At the same time, the level of Lake Victoria was altered by the tilting of the basin, so that its north end was considerably deepened, until overflow took place at what is now the Ripon outlet. Thus, the long succession of 100 ft. deposits up the Kagera valley ended with the Middle Tumbian and Middle Levalloisian clays.

The archaeological story is picked up again at Jinja, and seems to carry on the culture sequence without a break. In this part of the lake basin there are evident signs of considerable flooding, due to the tilt, and the drowning of old valleys and bays was of such extent that they are still somewhat submerged, in spite of the establishment of the Ripon outlet, soon after the maximum drowning of the region.

In some of the shore deposits of this maximum rise, we found a well-developed Late Kafuan industry, strikingly similar to that of the Kafu "flats" terrace. The majority of the tools were not water-rolled, and so must post-date the high-water level, but a few others, of the same type, are rolled and suggest that the whole industry was roughly contemporary with this rise of the lake. Probably, as soon as the Ripon outlet was cut, the people followed the retreating waters, whose previous shore deposits then became a land surface, in which the tools were embedded.

In the Kafu valley, we investigated numerous sites in brickearth and lateritic rubble on the 50 ft. terrace, as well as sites in the "flats" terrace. The tools from the latter are all rolled, some only slightly, others heavily, so that some might be derived from the 50 ft. terrace, but those in the brickearth and rubble are all absolutely fresh, though stained. They occur in definite horizons in the brickearth and this fact and the presence of much debris and many hammerstones, point clearly to their close association with home-sites. Both the open stations and "flats" series are of the same, very advanced facies, except for the very rolled, dubious examples from the terrace. We obtained about 575 artifacts from the camp sites, and about 450 from the terrace, so we had an excellent series for study and comparison with other Kafuan material from older or contemporary deposits.

Typology

For the first time, in dealing with this culture, it is possible to regard the tools as belonging to more or less definite classes, according to type and supposed usage. They comprise chisel-ended tools, points, end- and side-scrapers, small, multi-flaked cores, flakes, spherical sling-stones (?) and hammer-stones. The only use ever made of the flakes appears to have been cutting, for they never show signs of being retouched into scrapers, etc. Frequently, the tools are double-edged and some are flaked on both faces, and, of these, a number of the points somewhat resemble small *bifaces*. It is probably this feature of the Late Kafuan that led to its originally being described as possessing crude hand-axes, and to the assumption that the culture was the forerunner of the "Sangoan-Chelleo-Acheulean".[1]

At Jinja, the Late Kafuan tools occurred in a small home-site, only a few yards wide, for, on either side, the deposit was barren, as far as we could see. Excavation here was very slow and laborious, since we had to blast and then break up the laterite blocks with hammers. Naturally, some of the implements firmly embedded in the ironstone were broken in the process, but we were able to collect quite a large series, and these, when compared with the Kafu "flats" material, were seen to belong unmistakably to the same stage. All the same types were present, including small, spherical sling-stones, which were more numerous here than at the Kafu sites.

Until the Late Kafuan was studied, no resemblance was apparent between this culture and any other in Africa, partly because we were, at first, so uncertain about its age. When the comparative study of all the material was made, however, it was clear that the Late Kafuan bears some resemblance to the Mumbwan of Northern Rhodesia, with which, in fact, Burkitt[2] had already equated some tools from Luzira, Uganda, which may belong to the Late Kafuan. The implements studied by Burkitt come from a rubble which offered no evidence of

---

[1] *Trans. Roy. Soc. S.A.* vol. XVII, part 4, 1929, p. 336, and *A.R.G.S.U.* 1932, p. 57.
[2] Wayland and Burkitt, Archaeological Discoveries at Luzira, *Man*, 1933, 29.

FIGURE 6

Kafuan Culture

Heavily rolled tools of the Lower Kafuan from the Muzizi 50 ft. $\pm$ terrace. With the exception of No. 6, they are all made on comparatively flat pebbles, a characteristic of this culture which led to the manufacture of large scraping or gouging tools rather than heavy, angular choppers, as in the Oldowan (*q.v.*).

$\frac{1}{2}$ size.

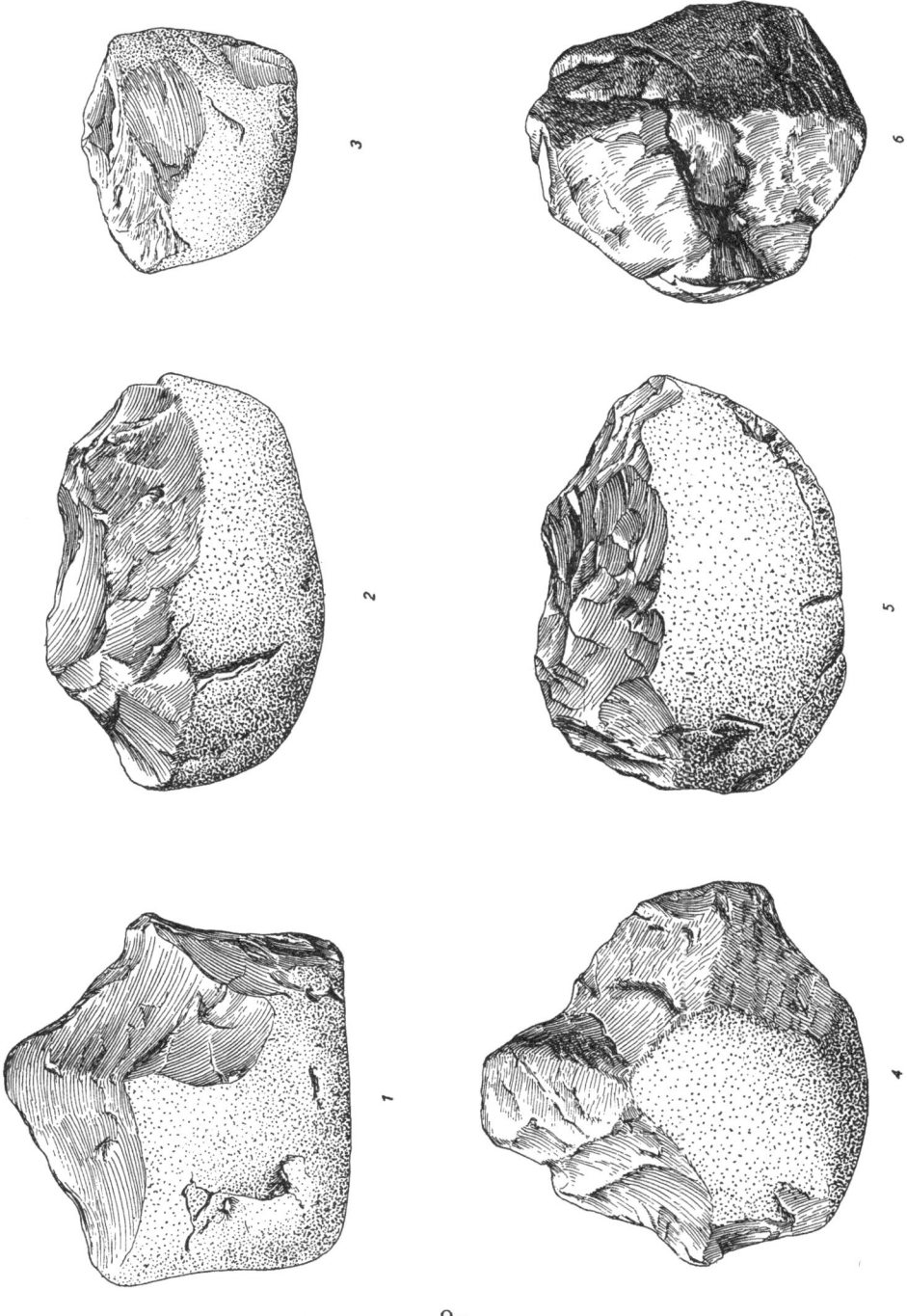

FIGURE 7

Kafuan Culture

Nos. 1–7: Rolled Middle Kafuan artifacts. 5 and 6 are from the 220 ft. ± beach at Mile 24¾, Nsongezi-Kagera Port Road; the others are from the Kafu 50 ft. ± terrace; quartz pebbles.

Nos. 8–12: Upper Kafuan artifacts from sites at the Kafu river. 8, 9, 11 and 12 are from laterite rubble sites on the same level as, and sometimes overlapping, the 50 ft. ± terrace gravels, and are fresh. 10 is a rolled example from the Kafu "flats" terrace; quartz pebbles.

$\frac{2}{3}$ scale.

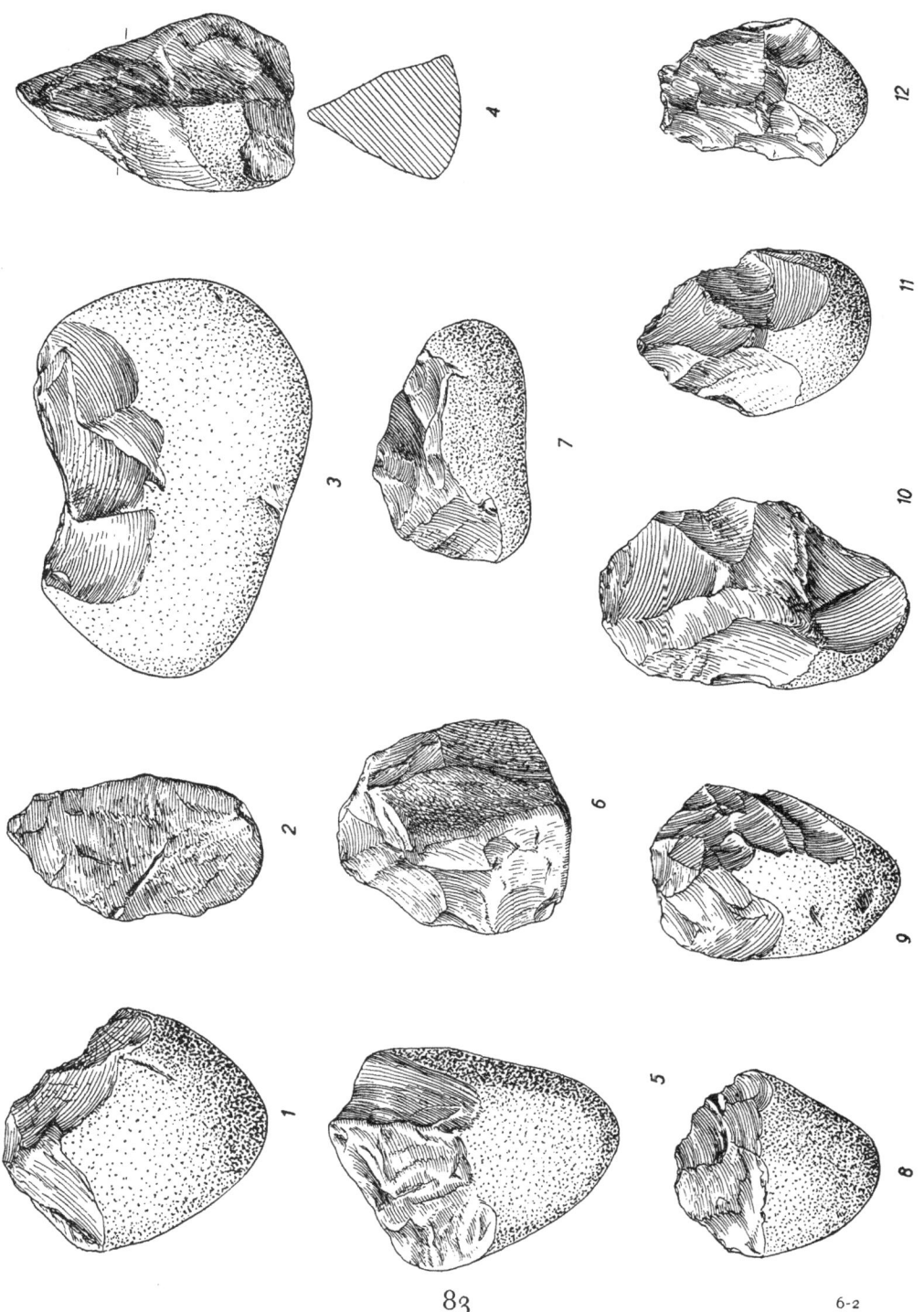

83

6-2

FIGURE 8

Kafuan Culture

Further examples of Upper Kafuan tools from sites at the Kafu river. 2, 3, 5, 8, 9, 10 and 11 are from horizons in brickearth on the same level as, and sometimes overlapping, the 50 ft. $\pm$ terrace gravels, and are all fresh. The rest are all slightly rolled examples found in the Kafu "flats" terrace gravels; quartz pebbles except 7 and 12 which are of brown chert.

$\frac{2}{3}$ scale.

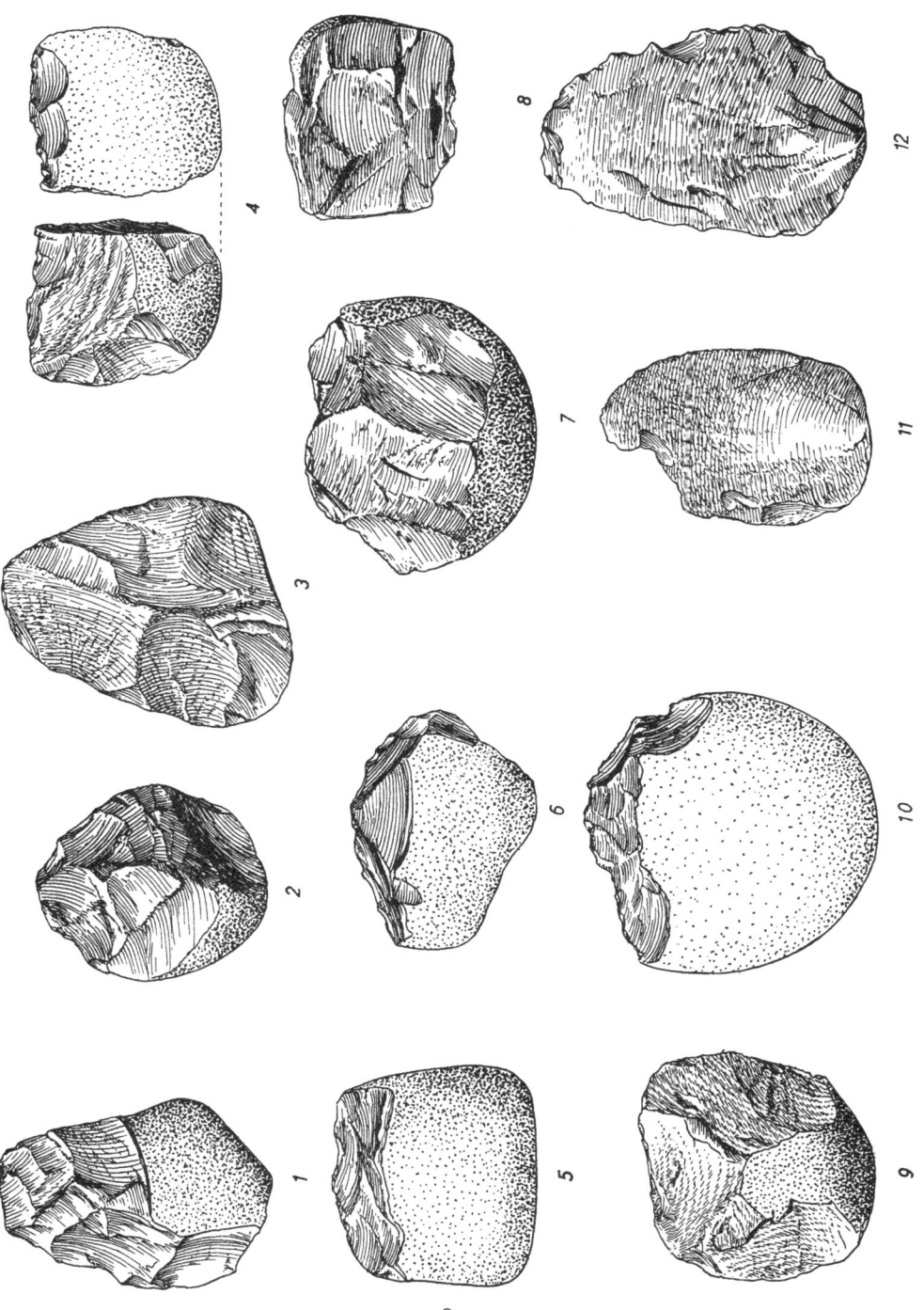

85

FIGURE 9

Kafuan Culture

Upper Kafuan tools from lateritised sediments at Bugungu, Lake Victoria (150 ft. $\pm$ rest-level). No. 1 is a slightly rolled example but the others are all fresh. 2 and 3 are small spherical sling-stones; 4, 8, 9 and 10 are flakes (bulbar sides below); 12 is a sort of hand-axe: it is flaked on both sides. It may belong to the Upper Tumbian culture. The others are typical Kafuan pebble scrapers, etc.; quartz.

$\frac{2}{3}$ scale.

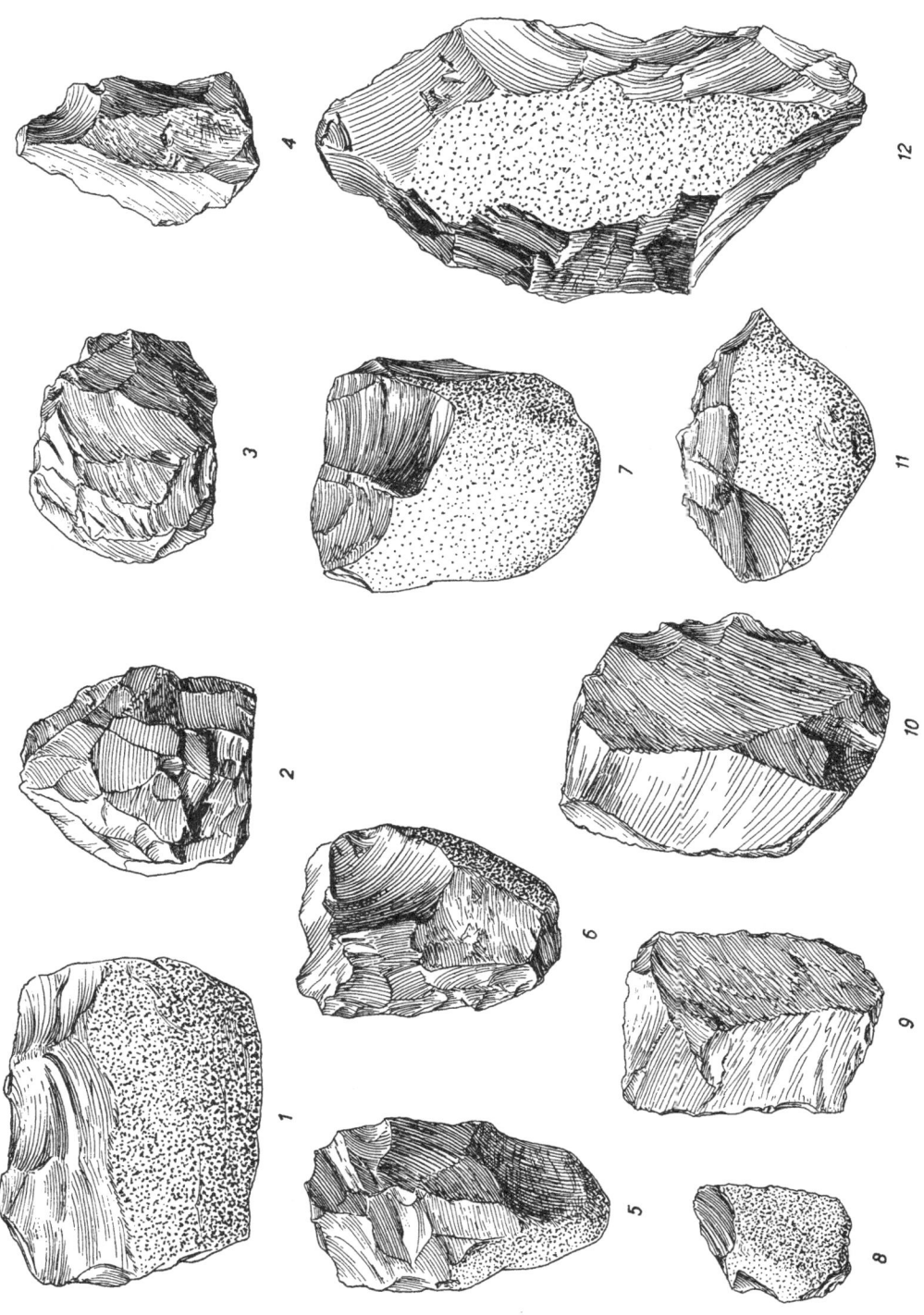

87

87

their age, except that they were later than a stage of Acheulean. They are not pebble tools, but made on chunks and it is likely that, wherever pebbles were not available, any other pieces of stone were used, and this may explain why, at Mumbwa, the majority of the implements are not made on pebbles. At Jinja, both pebbles and chunks were used as raw material. Except for the presence of true flake-scrapers at the type station, both the Mumbwa and Luzira series are very similar to the pure Late Kafuan, though they are quite possibly later and somewhat influenced by contact with Levalloisian. In fact the closest parallel to the Luzira tools is an industry that we found in a lateritic rubble, down in the Albert Rift, due north of Fort Portal, where, among typical Late Kafuan implements, there were a number of Levalloisian cores and prepared flakes. As they all bore traces of fairly recent entombment in solid laterite, and show similar states of preservation, they seem to belong to one stage, though it cannot be stated with absolute certainty.

Similar Mumbwan-plus-Levalloisian industries are, indeed, very common in most parts of Uganda, in a ubiquitous rubble below brick-earth, or in the latter, and having every appearance of being late, degenerate forms of the pure Late Kafuan.

SUMMARY

1. Wayland's view that the whole of the Kafuan is pre-Chellean in age is not upheld by either geological or archaeological evidence.

2. The earliest dated Kafuan is pre-Chellean, as demonstrated by the evidence from the Albertine Rift.

3. Middle Kafuan is probably about the same age as the Late Oldowan and Early-Middle Acheulean (Phase A) in the Kagera valley.

4. Late Kafuan is not older than Middle Tumbian and Middle Levalloisian, i.e. Upper Palaeolithic of Uganda.

5. The intense exclusiveness of the Kafuan right down to Upper Palaeolithic times suggests some marked racial character that rendered contact with other peoples more or less impossible; hence the isolated nature of various Kafuan industries except towards the close of its history, when it appears to have come into contact—perhaps only culturally—with the Levalloisian.

CHAPTER VII

The Uganda Oldowan

THIS culture was discovered in 1919, by Wayland, in the Muzizi valley. It was called Developed Kafuan or Muzizian, and subsequently equated with the Oldowan of Bed I, Oldoway, Tanganyika, in *R.R.R.E.M.U.* p. 336.

GEOLOGICAL DATING

In March 1933, the Royal Anthropological Institute's Conference met at Cambridge, where, among other matters concerning Early Man in East Africa, the archaeological sequence provided by the Oldoway discoveries was discussed. Soon afterwards[1] appeared the report of the Conference, in which the Archaeological Committee stated their conclusions, after examining the series of tools from Oldoway, as follows:

At Oldoway, in a continuous stratified deposit, which should henceforth rank as a standard section, a worked pebble industry in bed I is supplemented in the lower part of bed II by an early Chellean industry with coup de poing *and rostroid forms; and the pebble types persist for a while. There are indications of continuity and of a gradually evolving technique between the pebble industry and the Chellean technique.*

At various times since then, the same conclusion, that the Oldowan pebble culture evolves into a simple Chellean stage, has been stressed by Leakey,[2] and has been generally accepted.

In Uganda, however, there is strong evidence to show that the Oldowan, though definitely pre-Chellean in origin, did not evolve into the latter culture, but persisted into very much later, post-Chellean times. That conclusion was strengthened by an examination of the Oldoway type series in the Coryndon Museum, Nairobi, where I made the following notes:

[1] *Man*, 1933, 66.
[2] Leakey, *Adam's Ancestors*, 1934, p. 105, and *Stone Age Africa*, 1936, p. 41.

89

Pre-Chellean: *Bed I: six implements, four in lava, one in white quartz, one in grey chert (?). Three are made on pebbles, and one of these is large, elongated and almost certainly Chellean. The rest are of Oldowan type.*

Chellean I: *Base of Bed II: six implements, two in poor quality quartz, somewhat stained, the rest are in lava. Four are hand-axes, of which three are well shaped. Two are small chunks of Oldowan facies.*

It will be noted that, even in Bed I, tools of hand-axe type are present and that, in the base of Bed II, they are already well shaped. I feel very strongly that the acceptance of the evolution from Oldowan to Chellean was a mistaken conclusion, not supported by the typological or technical evidence of the tools themselves. One reason for this acceptance, probably, was the insistence that Oldoway is *the* East African Lower Palaeolithic succession, without taking into consideration either that it is, after all, only a single locality, or the possibility, now proved, I believe, of the independent local development of some industries, due to geographical or other causes. Thus, even on theoretical grounds alone, there is no reason why the disappearance of "pure" Oldowan at the type station should, necessarily, prove its disappearance elsewhere by "pure" Chellean times. Local impacts of culture are common, and these did not always at once lead to the absorption or suppression of one or other group, but, sometimes, to its dispersal or migration to other areas where it could continue its individual culture unchecked. Such, it seems to me, is all that we can believe to be probable in the case of the Oldowan. Very likely, further work will show that the Early Chellean was already in existence during the Bed I period and that it came into the area occupied by the Oldowan culture, only to drive this out. Further, there is no proof, at Oldoway, that the Oldowan tools found there represent the whole history of that culture.

It is not necessary to repeat the geological proof of the pre-Chellean age of the Lower Uganda Oldowan, which has already been given in detail in the previous chapter on the Kafuan. It may be briefly summarised as follows:

(1) There are two cultures in the Muzizi 50 ft. terrace and Epi-Kaiso Boulder gravels.

The Uganda Oldowan

(2) The first of these cultures is the more heavily rolled; typologically, it is distinct from the second, and has been given the name previously applied to tools of this class—Kafuan.

(3) The second group is much less heavily rolled than the first and is also typologically distinct from it. As recognised by Wayland, it is similar to the Oldowan and so is referred to that culture.

(4) On the Epi-Kaiso Boulder-Bed, heavily weathered, but unrolled, Chellean hand-axes have been found, made on boulders obtained from the bed.

Typology

The Lower Uganda Oldowan tools are large, clumsy choppers, made on pebbles and boulders of quartz and quartzite and of roughly spherical or cuboid shape. In most cases, largish flakes were detached by alternate blows, so that the intersection of several of the flake-scars formed a low, jagged chopping edge, while the spherical shape of the original stone was usually conserved. Another common form was produced by using a cuboid pebble, employing one flattish side as a striking platform and detaching flakes part or all the way round the edge. The result thus achieved was a sort of mammoth steep-scraper, but the edges were also tough and excellently suited for chopping purposes. Even the large implements are usually "handy" to grasp and wield. A few pointed tools and real scrapers also occur.

UPPER UGANDA OLDOWAN

We know of no Oldowan industries which can be proved to bridge the long gap between the Lower and Upper stages, though they must exist somewhere in East Africa, if not in Uganda.

The final appearance of this culture is with the Early-Middle Acheulean, Stage A industry, at Nsongezi, Kagera valley. For a time, we thought that these characteristic core-chopper tools were derived, as they bore so little apparent relation to the Acheulean. This idea seemed to receive support from the fact that the Oldowan implements from some sites round Nsongezi were clearly older than the Acheulean tools and had, in some cases, been reflaked. Subsequent work, however, showed that two periods of Acheulean exist at Nsongezi. The earlier

FIGURE 10

Oldowan Culture

Rolled tools of the Lower Oldowan from the Muzizi 50 ft. $\pm$ terrace, with the exception of No. 1, which was found in the Epi-Kaiso "Peneplain" Boulder Bed, Albert Rift Valley. The latter and 6 are unusually flat examples in a culture composed mainly of thick, spheroidal choppers. 6 itself is an interesting double-ended tool, perhaps intended for both boring and scraping uses.

$\frac{1}{2}$ scale.

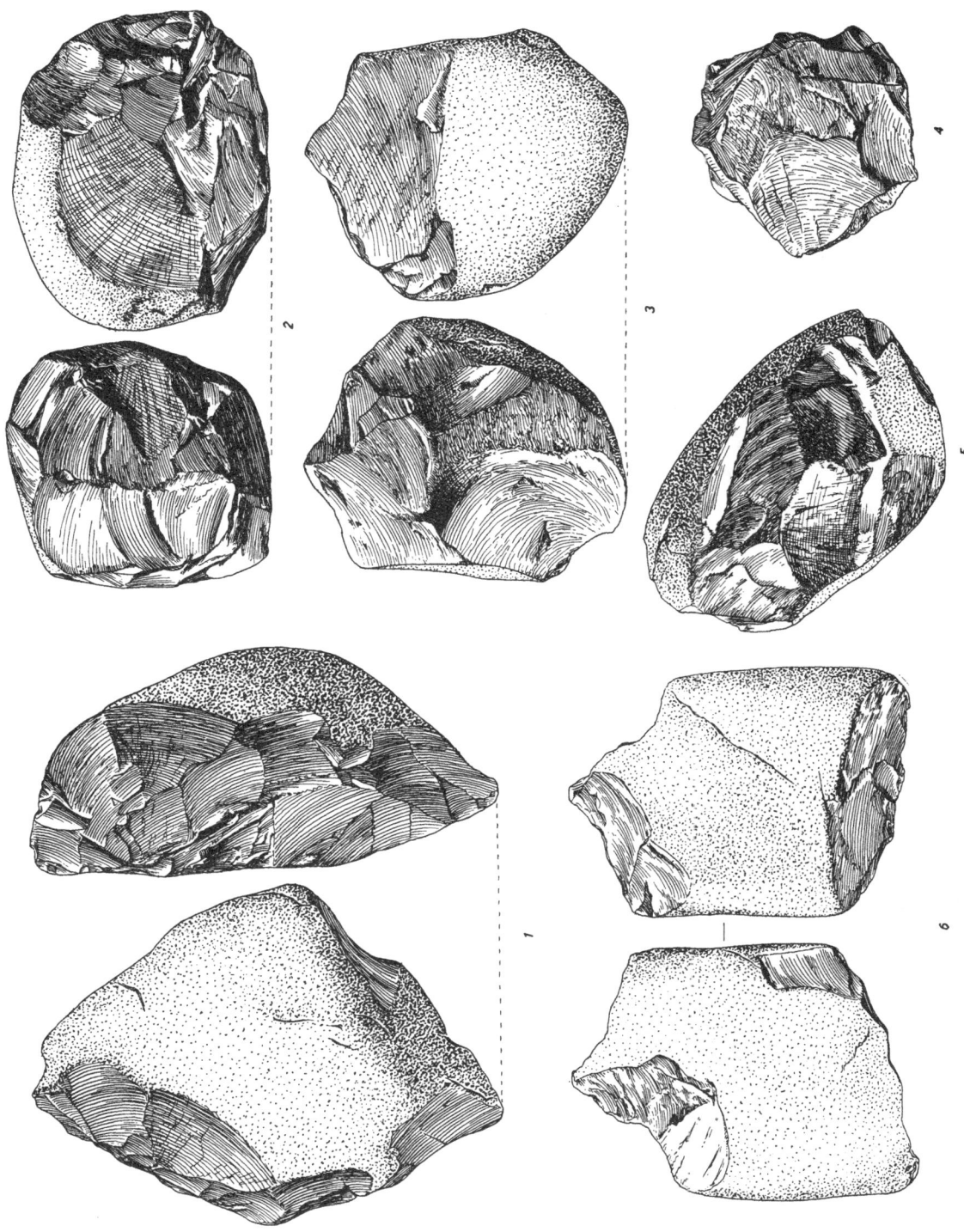

93

stage was found in a beach-gravel, in one instance, and in a shore deposit in another, and in both of these the obvious Acheulean tools were clearly of the same age as the Oldowan, exhibiting the same degree of rolling. Later, it was seen that the Acheulean of this stage was typologically earlier than that belonging to the arid rubble which Wayland had named the M-Horizon.

In the older Acheulean gravels (Stage A), Oldowan core-choppers formed 83% of the worked tools, Acheulean hand-axes 13% and cleavers 4%. At another site (see p. 67, fig. 1), near Jinja, some 190 miles away as the crow flies, a very similar gravel produced an industry very like that from Nsongezi. Near Jinja, the core-choppers formed 84% of the total and the Acheulean tools 16%, a striking resemblance to the Nsongezi figures.

Examination of the implements from the younger part of the M-Horizon, that is, Middle Acheulean, Stage B, showed that the number of Oldowan core-choppers had dropped to 45%, while the hand-axes and cleavers were in almost equal proportions, totalling 55% between them. Furthermore, the Stage B core-choppers are no longer of true Oldowan type, but considerably smaller and more advanced, and now form an essential part of the Acheulean industry.

Consideration of these facts has led to the conclusion that the Nsongezi and Jinja industries, Middle Acheulean, Stage A, represent a contact between pure Acheulean and a late phase of the Oldowan culture which gradually became absorbed until, by Stage B, the whole industry was Acheulean, though including modified Oldowan forms.

Although these facts seem clear, I would not go so far as to suggest that this contact was universal in its effect on the Oldowan culture, which may have continued to develop elsewhere for some time longer. Up to the present, in Uganda, however, there is no sign of any Oldowan industries later than that described above, and we may conclude that, locally, at least, it marked the culture's end.

Further aspects of this contact will be dealt with in the Middle Acheulean section of Chapter IX.

CHAPTER VIII

The Chellean

CHELLEAN tools had been discovered in Uganda by Wayland, but, apparently, had not been specifically recognised for some years and were included under the general heading Chelleo-Acheulean. Up to 1927, even "*the most obvious* coup de poing *and ovates*" were not regarded as representatives of the "Chelleo-Acheulean" culture, but were thought to belong to the Sangoan, which was considered as "*essentially le Moustier in facies*".[1] Apparently this was due to the discovery of hand-axes with scrapers and flakes of seemingly Mousterian affinities, at the Sango Hills type station, some years previously, all of which were thought to belong to one, non-Chelleo-Acheulean, culture. By 1929, however, Wayland had recognised that the Sangoan was not a specific culture, for he wrote:[2]

The Sangoan is a mixed culture, or rather, the result of a mixture of cultures of different ages, and lately the means of sorting out its elements has been provided by the discovery of deposits wherein these elements are separated...it would seem probable that pre-Chellean, Chelleo-Acheulean and Mousterian facies are all represented in the Sangoan portmanteau; but the general failure hitherto to discover these elements unmixed or insufficiently mixed to render them separable, led one, for the time being, to doubt whether definite Chellean (or Chelleo-Acheulean) and Mousterian culture facies had existed separately.

Wayland was undoubtedly right in recognising the "Sangoan" as a mixture, though, more recently, he seems to have returned to the view that it is a single entity, when he states:[3]

Leakey and Solomon had come to regard my Sangoan 'as partly Chellean, partly local Acheulean with a hint of Levalloisian'. They were not far from the truth; the culture is not a superficially mixed one, however....

Typology apart, there can be no possible doubt that the Sangoan is a mixture, though of much less significance than is implied in the view of

[1] *A.R.G.S.U.* 1927, p. 34. [2] *Sum. Prog. Geol. Surv. Ug.* 1929, p. 40.
[3] *R.R.R.E.M.U.* pp. 342 and 351.

Leakey and Solomon in the statement just quoted. Nor is it likely that Wayland's own belief in the Chellean, Acheulean and Mousterian (that is, Levalloisian) mixture was correct. A study of the physical condition of the implements immediately establishes the fact of mixture, but the successive stages of culture only begin with comparatively late Acheulean, while the bulk of the more typical "Sangoan" belongs to the Proto-Tumbian stage of culture, of post-Late Acheulean age. Still later phases are represented by Middle Tumbian *bifaces*, while the Levalloisian element was a contemporary, but separate culture, dating from the time of, or shortly before, the Proto-Tumbian. It was probably the extreme crudity and great size of some of the tools of the latter industry that led to a false idea of their age and to comparison with the true Chellean, of which, in my opinion, there is not a trace in the "Sangoan".

By 1933,[1] however, the presence of true Chellean seems to have been recognised, together with its similarity to the European and South African branches, although, in Wayland's paper of the following year,[2] no mention is made of it as a specific culture, and the whole hand-axe group (with the exception of the "Sangoan") is again referred to as Chelleo-Acheulean.

Judging by our own experience, the difficulty of recognising the Chellean as a distinct, pre-Acheulean culture was probably largely due to its comparative rarity and also to the fact that it usually occurs in rubbles in which other later assemblages are often also present. We never found it in bedded deposits of contemporary date, but it was abundant in a thick rubble in the Kagera valley. This fact is of climatic significance, as will be explained shortly.

TORO-SEMLIKI AREA

The geological succession established in this area (the south-east corner of the Albert Rift) has already been dealt with in Chapters III and VI. We have shown how the Epi-Kaiso lake floor was covered by a boulder-bed containing Lower Kafuan and Lower Oldowan tools and that, later, this boulder-bed became a land surface, used by Chellean and later peoples as a source of raw material.

[1] *A.R.G.S.U.* 1933, p. 57.   [2] *R.R.R.E.M.U.* 1934.

The Chellean

We were unable to determine, for certain, the cause that led to the exposure of the Epi-Kaiso Peneplain Boulder-Bed, but, as the deposit shows no sign of a gradual decrease in grade size or further laying of fine sediments above it, it seems likely that the fall of the lake was sudden. The Kisegi-Kaiso-Epi-Kaiso sedimentary series here forms a plateau, now truncated, on its western edge, by a steep scarp, so it is quite probable that faulting or sharp flexing,[1] which led to the formation of the present lake basin, took place shortly after the laying of the Peneplain Boulder-Bed. Near Nyambirizi, the Epi-Kaiso plateau lies nearly 1000 ft. above the present lake level (see p. 24, fig. 1).

There is nothing to indicate what length of time elapsed between the formation of the Peneplain Boulder-Bed and the first appearance of the Early Chellean, but the evidence from the Kagera valley suggests that it must have been long enough for a considerable change of climate to have taken place.

KAGERA VALLEY

In this valley, the road from Nsongezi to Kagera Port runs, for the most part, along the foot of the hills that enclose the valley. A number of pits have been dug for road metal by road-maintenance gangs, in the rubble accumulations at the foot of the hills, generally above the road. These pediment rubbles consist chiefly of angular blocks of quartzite that have weathered out of the hillsides and slowly accumulated on the slopes and at their feet. According to both Wayland and Solomon, independently, their formation depended upon the absence of any but the slightest rainfall, otherwise, sufficient vegetation would have been present to hold up the downward movement of individual stones. This fact is well attested to-day, when only earths, and no such rubbles, are forming. So it was obvious that these rubbles were of importance to the question of past climates in general, and for the dating of the stone tools in them, in particular.

In the Kagera valley, we found that there are two hillside rubbles which contain tools of various periods, covering much of the Palaeolithic sequence. During moister periods, it may be assumed that tools

---

[1] Even this movement, however, must not be thought comparable with the great faults that formed the Rift scarps in much earlier times.

dropped on the hillsides were simply covered temporarily by earth, and that subsequent erosion swept them to lower levels in the course of time. There is little or no trace of such "fossil" earths on these hillsides to-day, but they have been preserved in more favoured places, such as short, lateral valleys that were not subjected to much vertical erosion. An instance of this is Wayland's hanging valley, near Nsongezi, where he dug through 65 ft. of alternating rubble and clayey deposits before reaching bedrock.

In practice it was very difficult to differentiate between successive rubbles on the hillsides except by their contained artifacts. After many sites had been examined it became clear that the Older Rubble is very much compacted and contains nothing but Chellean tools and flakes, while the Younger Rubble is loosely formed, contains a good deal of earthy material and, above all, is locally prolific in Acheulean and later assemblages. Further, the implements from the Older Rubble are extremely heavily weathered, both mechanically and chemically, while those from the Younger are much less weathered (though degrees of wear are observable, corresponding with the relative ages of the tools) and not chemically disintegrated at all. At a few sites we saw an actual earthy division between the two, which provided further proof of their separation.

Wayland's 65 ft. pit in the hanging valley, near Nsongezi,[1] provided clear evidence of the alternation of dry periods of rubbling and more moist phases when clays or clayey earths were formed. As a number of these deposits contain stone tools, some of which have been called Chellean, it is necessary to re-examine this evidence here, from the point of view of this culture.

Briefly, the succession is as follows, the information being taken from Wayland's diagram in the paper cited above, and the section reading from the top downwards:

F. Red argillaceous deposits; Acheulean tools; about 5½ ft.; pebbles at base.

E. Rock rubble; a few crude Chellean tools; about 11 ft.

D. Micaceous clay; about 7½ ft.; a few pebbles at base.

[1] *R.R.R.E.M.U.* p. 345.

C. Rock rubble; a few crude artifacts; about 30½ ft.

B. Red clay; about 5½ ft.; inconstant pebbles at base.

A. Rock rubble; perhaps a few crude artifacts.

Leaving the lowest rubble (A) out of account for the moment, the reported Chellean tools of Rubble E would seem, at first sight, to contradict the existence of our second rubble, containing Acheulean, etc. It is inconceivable, however, that any dry period capable of causing rubbling in far less favourable places than this, should not have done so here; in fact, it is more likely that, in such a comparatively narrow space as this, any such rubbles as formed would have been thicker than those of the same age on the hillsides round the main valley.

The apparent anomaly is explained, however, when the supposedly Chellean age of Rubble E is examined critically. All these deposits have been equated by Wayland with various climatic phases of his pluvial scheme, with the result that Rubble E is equated with Intrapluvial 2. The latter is the dry period responsible, in Wayland's opinion, for the M-Horizon of the Kagera 100 ft. terrace, which he has correlated with Bed III, at Oldoway, that is, in his view, *Early Acheulean*. Our work, on the other hand, has shown that the M-Horizon material is *Middle* Acheulean, and comparable to Oldoway Acheulean 3–4. So it is difficult to see how the Rubble E tools can be Chellean, unless they have weathered out of an older deposit. Considering the great thickness of Wayland's Rubble C and our Older (Chellean) Rubble, it seems likely that these two are the same and that his Rubble E is the equivalent, at least in part, of our Younger Rubble. Furthermore, both Chellean and Lower Acheulean only occur sporadically in the valley and both contain some very crude tools, so these facts may explain both the apparent absence of true Chellean from Rubble C, and the crudity of those from Rubble E.

The lowest rubble (A) does not call for comment here, except that, as we never bottomed the Older Rubble in the main valley, we do not know if Rubble A is represented there, or what tools, if any, it contains. Presumably, it was formed during some pre-Chellean dry period, but it may be of any age—even pre-Pleistocene—since there is nothing to date the original cutting of the hanging-valley channel.

CLIMATE

All the evidence so far examined indicates that the wet period of the Epi-Kaiso Boulder-Bed and the Muzizi 50 ft. terrace gravels was followed, after an unknown lapse of time, by a long dry one, to which the Older Rubble belongs. Wherever a good section was visible, this rubble appeared to be of considerable thickness. As far as we know, it contains only one culture, the Chellean, and, judging by Oldoway standards, a fairly early stage of it, though it is not so crude or small as the oldest stage at Oldoway. It is only possible to conjecture why no later stages are present in the rubble, but a possible reason may be that the country was only sparsely inhabited in the first place, and that, after a comparatively short time, it was abandoned in favour of areas with a more favourable climate or water supply.

Considering that the present rainfall of this part of Uganda is far from heavy, and yet there is a certain amount of vegetation, it seems almost certain that, during the periods of rubble formation, the climate must have been very dry indeed, to account for the almost total extinction of binding vegetation whose presence would have prevented the formation of rubbles.

TYPOLOGY

The Lower Chellean from the Toro-Semliki area is made in a variety of rocks and always from boulders obtained from the Epi-Kaiso Peneplain Boulder-Bed. As the best Chellean collections were made in the Kagera valley, however, I shall not give a detailed description of the Toro-Semliki specimens, all of which types (in so far as there are types) are duplicated in the Kagera valley material.

The tools from the Kagera valley Older Rubble are usually made on large pieces of grey quartzite, obtained from the local outcrops and, doubtless, from the angular rubble bed itself. They are large, very heavily weathered and variously stained.

As a rule, the hand-axes are humpbacked and asymmetrical and not always of very obvious *coup de poing* form. They are seldom worked all over or all the way round. The usual method of manufacture was to select a lump of rock of suitable size and then to strike off a few big flakes near one end, producing a rough, pointed tool. Another method was that followed at many different periods by makers of hand-axes. I have never seen this technique described, but it must have been a common one in Africa wherever tabular quartzite occurred. This method consisted of selecting a piece of stone with two natural cleavage or joint planes approximately at right-angles to one another. Then, using the outer edges of these flat faces as striking platforms, flakes were detached all round the periphery of the stone, producing a hand-axe, pointed at one or both ends, with an inverted triangular section whose upper face is fairly flat and worked all over, while the lower face is steeply keeled, where the two original, unworked, right-angled planes converge. I have called these hand-axes the base-keeled type.[1]

Other types of core tools include rostroid hand-axes and large choppers.

It seems that the Uganda Chellean people, like the Early Stellenbosch folk of South Africa, used flakes as well as hand-axes, for we found quantities of the former, far too large to have been simply the waste material detached in making core tools. Indeed, for some time, these flakes (which we sometimes found by themselves) were regarded as a distinct, early flake culture, and were given the name "Uganda Cromerian".[2] While not denying the possibility that some of them *may* belong to such a culture, I feel that, at present, it is safer to regard them as essentially part of the Chellean, until future work can decide. In any case, the use of such a local term as "Cromerian" to describe an extremely simple flake industry in Africa is open to grave objections which I did not realise when the term was first applied to the Kagera valley material. The name was given because, in the majority of cases, the flakes did not have markedly inclined striking platforms and prominent bulbs of "Clactonian" type, but were usually flattish, with small

[1] In places like the Sango Hills, where almost all the raw material tends to break up into rectangular or tabular blocks, this method was very extensively used during the Proto-Tumbian occupation of the area, in Upper Palaeolithic times. Naturally, the appearance of such tools is extremely coarse and primitive, and this probably explains why several other workers have thought that the "Sangoan" was an earlier culture than it actually is.

[2] *Man*, 53, 1936.

PLATE V

Chellean Culture

Fig. 1. A large hand-axe from the Kagera valley Older Rubble; quartzite, heavily weathered.

Figs. 2 and 3. Two hand-axes from the surface of the Epi-Kaiso "Peneplain" Boulder-Bed, Albert Rift Valley; made on quartzite boulders, heavily weathered. All these specimens are repeated on pp. 109, 111 and 117.

Approx. $\frac{1}{2}$ scale.

PLATE VI

Chellean Culture

Fig. 1. Two views of a rostroid hand-axe from the Kagera valley Older Rubble; quartzite, heavily weathered.

Fig. 2. Two views of a hand-axe from the same deposit; quartzite, heavily weathered. Both these tools are repeated on pp. 113 and 115.

Approx. $\frac{1}{2}$ scale.

I

2

105

PLATE VII

Chellean Culture

Large flakes from the Kagera valley Older Rubble near Nsongezi; quartzite, heavily weathered.

Approx. $\frac{1}{2}$ scale.

1

2

FIGURE 11

Chellean Culture

A very large hand-axe from the Kagera valley Older Rubble; quartzite, heavily weathered.

$\frac{1}{2}$ scale.

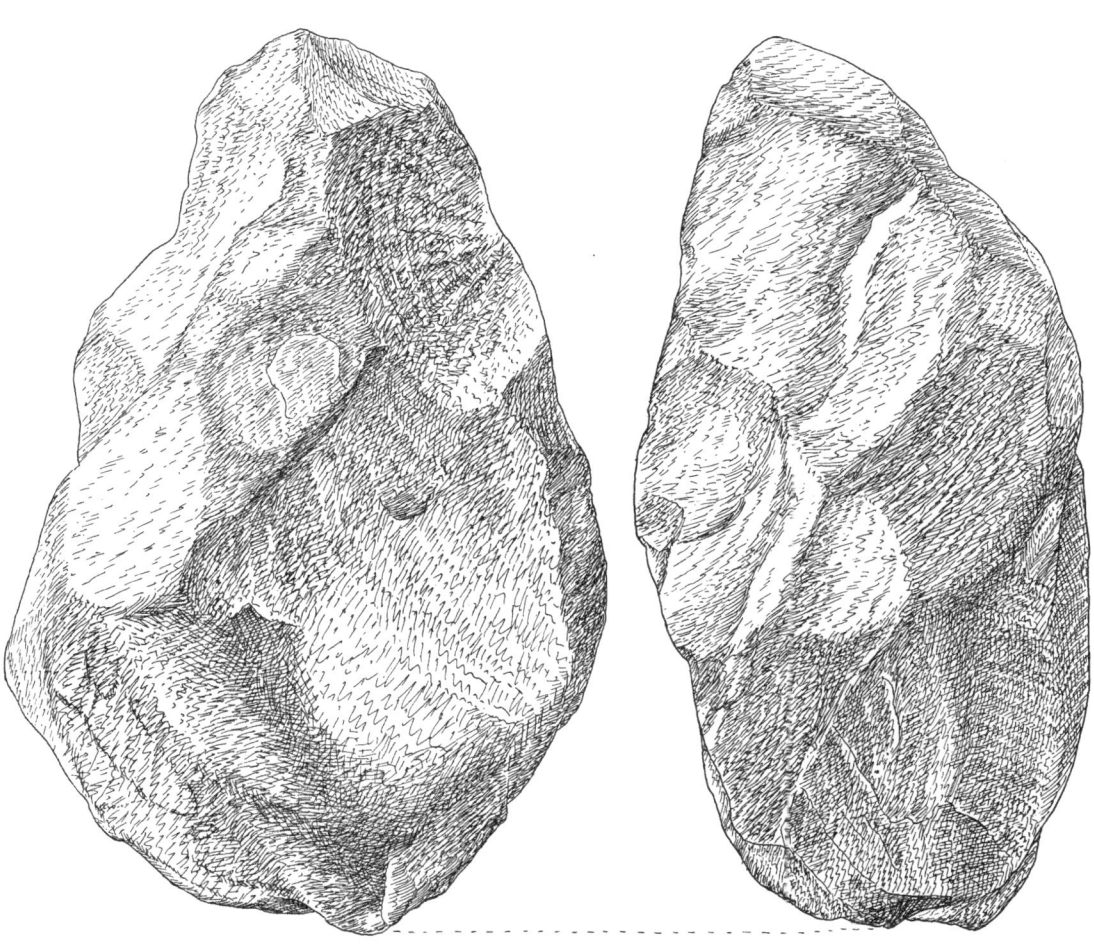

FIGURE 12

Chellean Culture

Two hand-axes from the surface of the Epi-Kaiso "Peneplain" Boulder-Bed, Toro-Semliki area, Albert Rift Valley; quartzite boulders, heavily weathered.

$\frac{1}{2}$ scale.

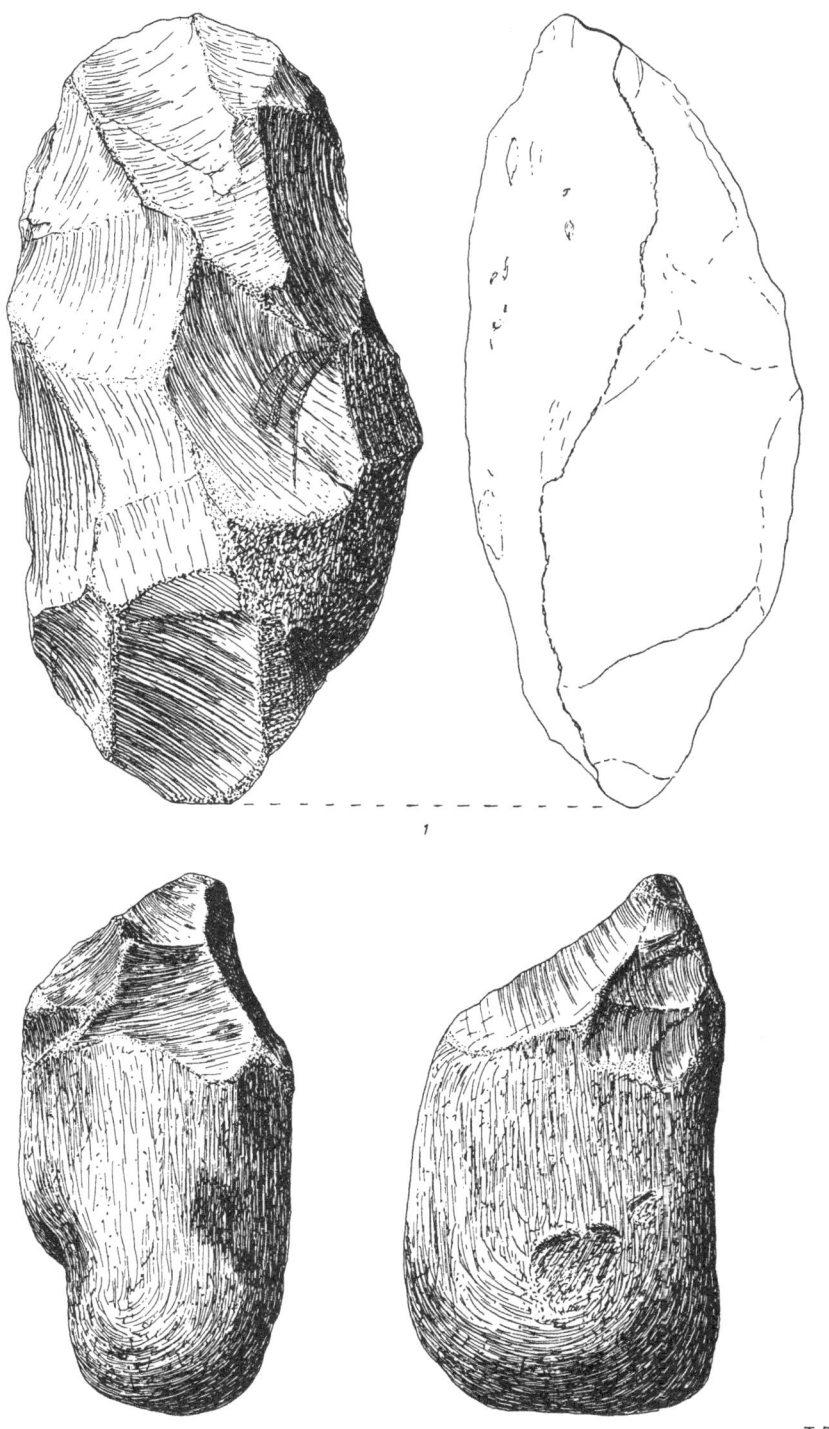

1

2

T.P.O'B

III

FIGURE 13

Chellean Culture

Two hand-axes from the Kagera valley Older Rubble. No. 1 is a typical "rostro-carinate" of the East Anglian tradition; quartzite, heavily weathered.

$\frac{1}{2}$ scale.

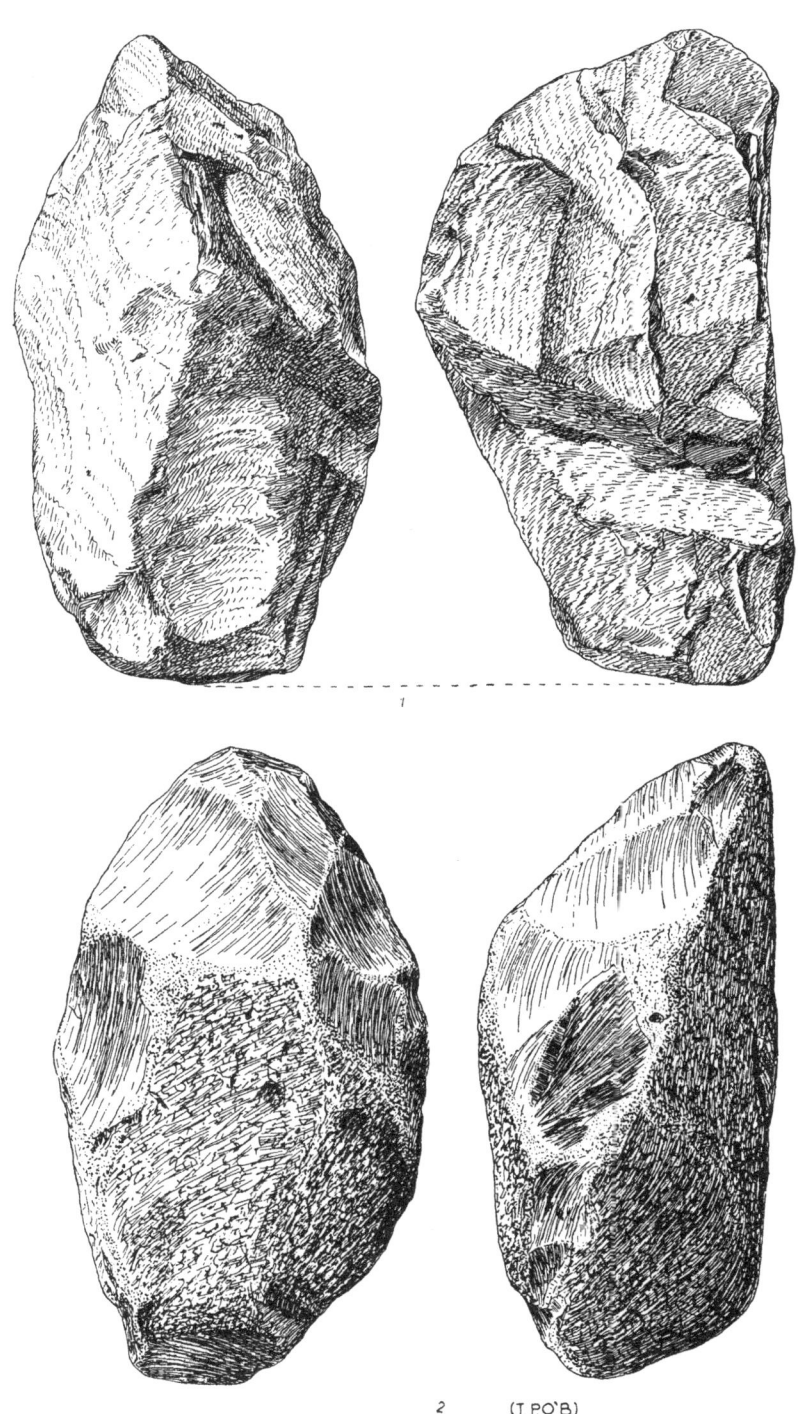

1

2   (T. P. O'B)

FIGURE 14

Chellean Culture

Two hand-axes from the Kagera valley Older Rubble; quartzite, heavily weathered
½ scale.

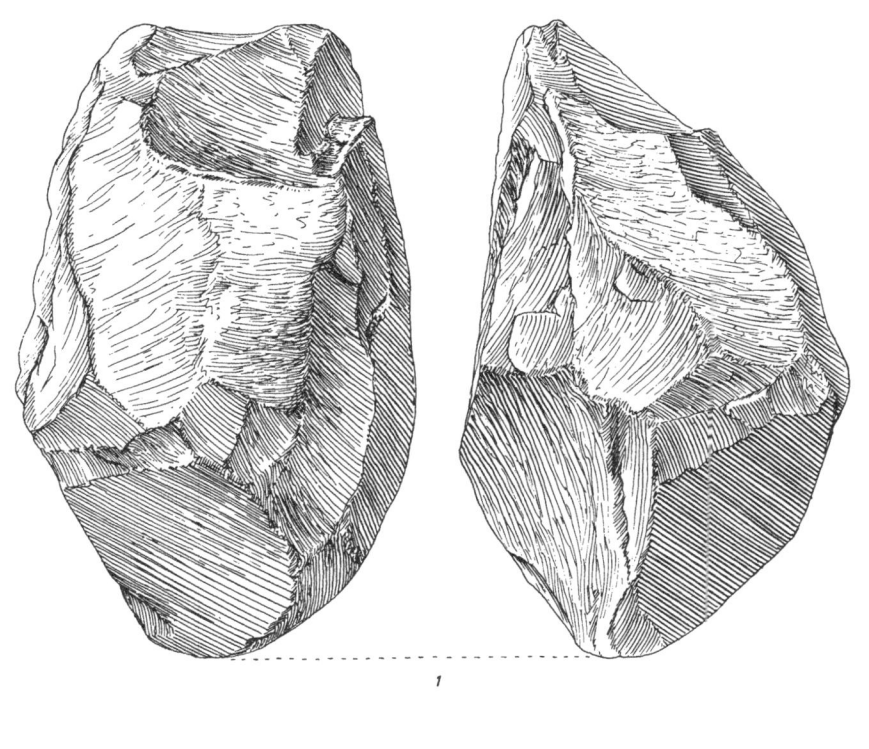

1

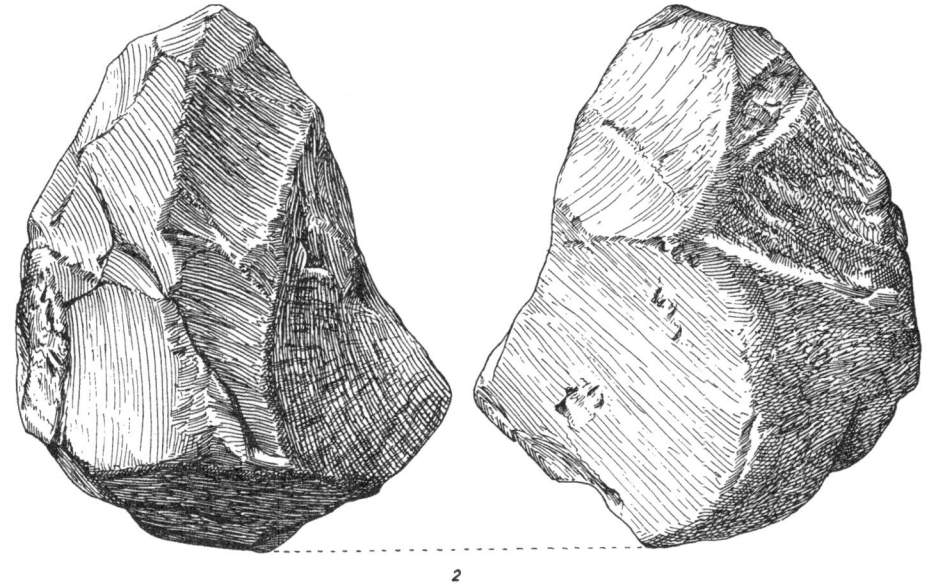

2

115

8-2

FIGURE 15

Chellean Culture

Nos. 1 and 2: Two base-keeled Chellean hand-axes from the Kagera valley Older Rubble; quartzite, heavily weathered.

No. 3: A hand-axe from the surface of the Epi-Kaiso "Peneplain" Boulder-Bed, Toro-Semliki area, Albert Rift Valley; quartzite (?) boulder, heavily weathered.

½ scale.

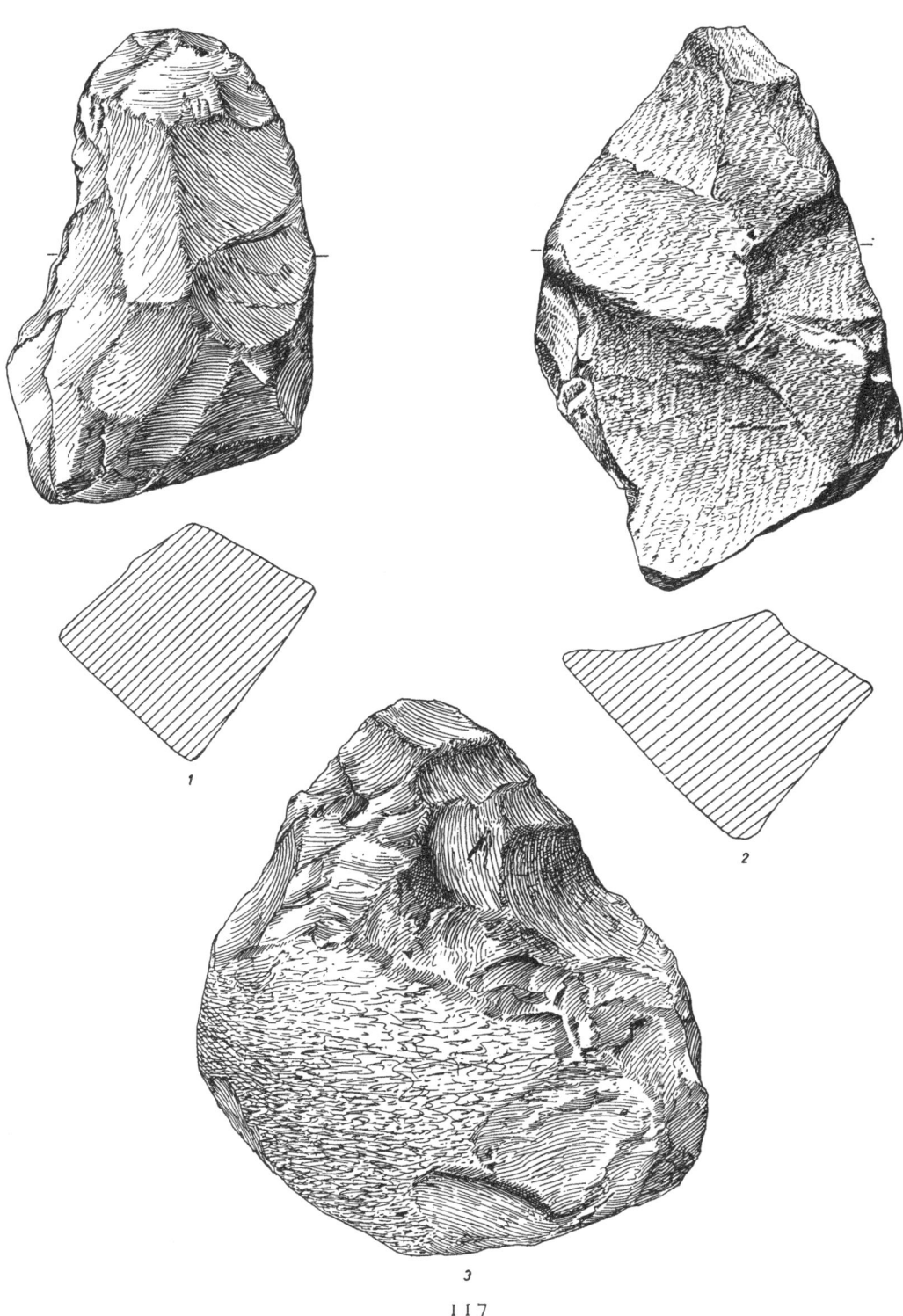

1

2

3

117

bulbs. In a few, rare cases, the flakes have been retouched into scraping and chopping tools.[1]

Whether these flakes belong to a distinct culture, or to the Chellean, as I, personally, feel that they do, there is every reason to think that they were contemporary with the obvious Chellean tools, for both are alike in their states of preservation. It is also to be noted that large flakes are common in all the later, Younger Rubble assemblages, where they are even more certainly a part of the various hand-axe groups.

[1] Our Lower Chellean is strikingly similar to the Early Stellenbosch of South Africa and it is to be noted that, there, core and flake tools occur in almost equal proportions, according to Professor van Reit Lowe, in the *S.A.J. Sci.* 29 October 1932. In fact, the Early Stellenbosch is sometimes referred to as "Chelles plus Clacton", though the implication of cultural mixture is not one that can be proved. It would appear that, at a very early stage, the makers of large core tools discovered the advantage of using large flakes as well, and the best method for detaching these (indeed, almost the only method) was the use of the "Clacton" technique, that is, swinging the core down on to a fixed anvil-stone. This simple but effective technique may easily have been discovered independently at various times, by different peoples.

CHAPTER IX

The Acheulean

As far as I know, the first attempt at subdividing the Uganda "Chelleo-Acheulean" complex was made by Wayland after examining the Oldoway succession. In the account of his visit and the accompanying table,[1] he equated the M-Horizon of the Kagera 100 ft. terrace with Oldoway, Bed III, the reddening of which he regarded as evidence of a dry oscillation like that of the M-Horizon. This correlation made part,[2] at least, of the M-Horizon material Lower Acheulean, for Bed III contains only a single early stage of this culture. For a long time we were obliged to follow this dating of the M-Horizon material, although we felt that the advanced technique of some of it did not agree with such an early date. It was not until we left Uganda and studied the Oldoway series in the Coryndon Museum, Nairobi, that proof was forthcoming, when we saw that the Bed III material was much cruder than that of the M-Horizon, which only forms one single industry, though with two substages. In many ways, particularly in the number of well-shaped cleavers and the beauty of the flaking, the M-Horizon material is very like Oldoway Acheulean Stage 3–4, and so is of Middle Acheulean date.

LOWER ACHEULEAN

When studying the rubble deposits in the Kagera valley, we had found that the Younger Rubble contained a variety of tools which proved to be divisible according to age, both by typology and state of preservation. The oldest of these comprised a small group of rather primitive hand-axes which were obviously older than the M-Horizon material and which, because of the "Lower Acheulean" age of the latter, we were obliged to regard as Upper Chellean.

As soon as the Middle Acheulean age of the M-Horizon was estab-

---

[1] *R.R.R.E.M.U.* p. 343 and table facing p. 344; *A.R.G.S.* 1932, p. 13.
[2] Wayland is inclined to the view that there may be several stages of culture in the M-Horizon (see Appendix A, note on the M-Horizon, T. P. O'B.).

lished, however, the nomenclature of the oldest group of tools from the Younger Rubble had to be reconsidered. In view of the prevalent belief that the Acheulean is the direct descendant of the Chellean, it would not, perhaps, have mattered much whether we called this series Upper Chellean or Lower Acheulean, but it was evident that some great gulf lay between it and the Uganda Chellean. Though still crude the industry is nearer to the Acheulean than to the earlier culture and, until the gaps in the latter's succession are filled, it is better to refer this group from the Younger Rubble to the Lower Acheulean.

Unfortunately, the Younger Rubble is not nearly such a compact and "fossil" deposit as the Older, Chellean Rubble. It is clear, from the admixture of several later cultures and industries, either that it has been much disturbed and resorted by erosion and, possibly, by man, or that the whole deposit continued to form for a long period, beginning with the Lower Acheulean and ending with the Middle Tumbian and Levalloisian—the latest tools to be found in it. Possibly, a combination of both circumstances occurred. For these reasons we cannot say for certain whether the Lower Acheulean tools became incorporated in the deposit during a dry climate (as in the case of the Chellean), or whether they are derived from older, "wet" soils which were eroded and incorporated in the rubble at a later date. The likelihood of the second alternative is suggested by the almost invariable absence of any earthy division between the Older and Younger Rubbles and the presence of a good deal of earthy material in the latter. In very favoured places there is a distinct break between the two, which has been interpreted as indicative of a moister climate than that prevailing during the rubble formation; it is possible that the Lower Acheulean belongs to this moist phase, although, so far, it has always been found in a rubble, and it is not represented in the lacustrine or riverine beds in the valley. In the present state of our knowledge, it is impossible to go further than the statement of these various possibilities (see p. 49, fig. 2 and p. 67, fig. 2).

Typology

I have already referred to the fact that, by some standards, the Lower Uganda Acheulean might be classed as Chellean, in view of its general crudity and lack of specialisation. The reasons why I cannot so class it are, first, that true, biconvex hand-axes occur, chipped all round and, secondly, that the chipping on the faces of the tools is much more extensive than in the Uganda Chellean, indicating a great advance in technique. In the Chellean the implements are essentially only roughly pointed lumps of stone, but the Lower Acheulean tools are edge-trimmed, as well as pointed and are sometimes worked nearly all over.

The only other important feature of this industry is the continuance of the use of the base-keeled technique, first seen in the Chellean. Indeed, most of the tools are keeled either on the dorsal or ventral face, according as to whether one or two faces of a triangular sectioned piece of stone were flaked. In each case we regard the worked surface as the upper, or dorsal face. If only one of the three sides is flaked, the two plain surfaces, intersecting at a keel, form the ventral face, but, if two sides are flaked, the intervening keel occurs on the dorsal face, with the worked areas on either side of it, and the plain face on the ventral side, below.

In almost all cases of hand-axes of the base-keeled type, laying the instrument down on either of the plain faces results in marked asymmetry, because, while the plain sides, with their intervening keel, are natural surfaces with a straight edge, the worked face tends to be rounded through having been flaked into a pointed end or ends. To make these tools symmetrical when viewed from above they must be held either with the keel pointing upwards or downwards. The first position usually results in the pointed ends curving up towards the operator—not a convenient position for use, so it appears that the keel was intended to be on the under side, hence my calling them base-keeled.

The fact that this technique is to be found throughout the hand-axe cultures of Uganda (except the Middle Acheulean Stage B, where the majority of the tools were made on flakes) wherever tabular quartzite was employed, shows clearly, I think, that it was no more than an adaptation for the production of hand-axes out of intractable material. Wherever tools were made on flakes, it was never used.

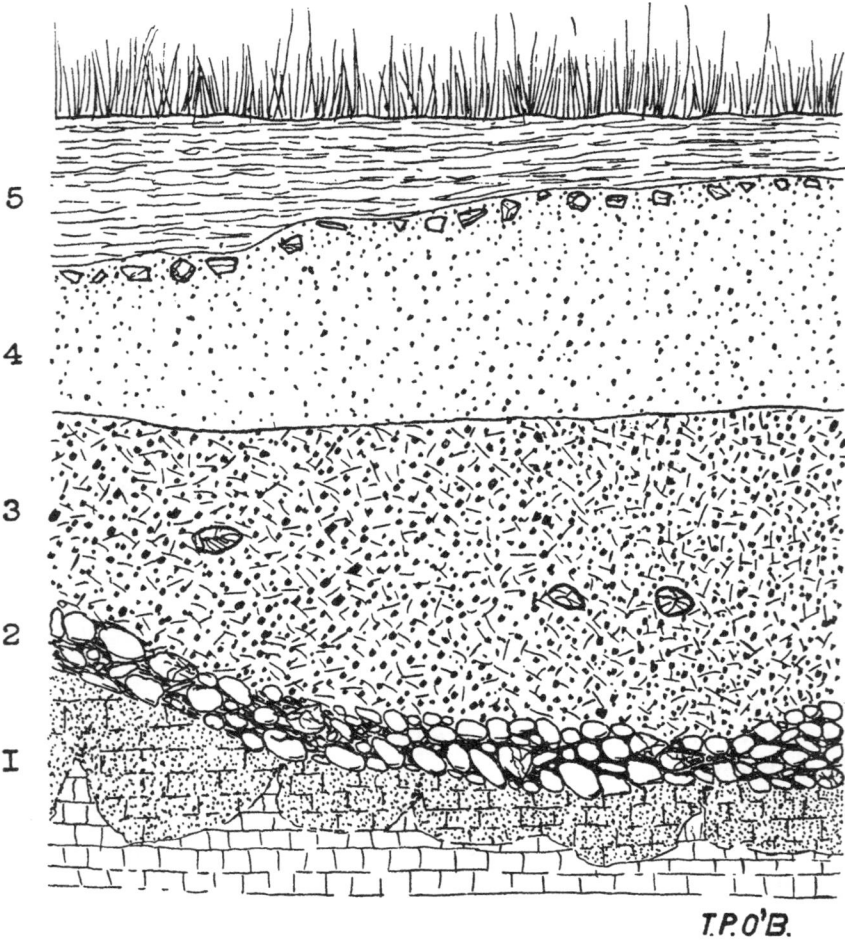

T.P.O'B.

FIG. 16. Section at pit near Jinja showing basal gravel
containing Early-Middle Acheulean industry.

5. Alluvium; 1–2 ft.
4. Fine laterite rubble with Levalloisian tools at top; 2–3 ft.
3. Pisolitic laterite with rare Levalloisian tools; 2–4 ft.
2. Eroded gravel bed, very uneven, clayey matrix, Early-Middle Acheulean and Late Oldowan tools; to 2 ft.
1. Rotted bedrock.

The Acheulean

EARLY-MIDDLE ACHEULEAN

The first Early-Middle Acheulean site[1] that we worked was at a pit dug for road metal, not far from Jinja. The implementiferous horizon was a gravel-bed below solid laterite (cf. p. 122). The majority of the artifacts were round or cuboid core-choppers and flakes, but a few hand-axes were also present. At that time we did not realise the true significance of this industry, as we had not investigated any other Lower Palaeolithic sites and the study of the more extensive deposits and industries at Nsongezi was still in the future. We thought that the Jinja industry was an Upper Oldowan stage, giving place to Chellean. Up to this time no one thought that the Oldowan continued later than the very early Chellean. All the Jinja tools were rolled and, though the degree of rolling varied somewhat, there was nothing to suggest that the hand-axes were later than the other implements, so we could only suppose that they were rather precociously well-developed Early Chellean. As we had no field knowledge then of the Uganda Chellean, we had no idea what its early phases might look like; moreover, there was little in the geology of the site to indicate the probable age of the basal gravel. The presence of a few Levalloisian tools in the overlying laterite and a still later stage of Levalloisian in a rubble above it, only suggested some long break between the gravel and the laterite phases. We now know that there was a break, but it was of less importance than we thought at first. Later work, at Nsongezi, showed that the hand-axes were Acheulean, as, indeed, their form had previously suggested.

Nsongezi

For several years before our visit, Wayland had carried out intensive studies in the Kagera valley, principally in the neighbourhood of Nsongezi. In the course of this work, which dealt mainly with the

---

[1] Reference has already been made, in Chapter VIII, to the Oldowan tools of the Early-Middle Acheulean at Nsongezi. In this chapter, only general remarks will be made about these, but it must be remembered that it is not possible to say, at present, whether they belong to an industry that was separate from, but in close contact with the Acheulean, or if it had already coalesced with the latter. For this reason, the Upper Oldowan and Early-Middle Acheulean of Nsongezi should be studied together, although, in this book, the descriptions of their distinctive tool-types are kept in the appropriate culture chapters, for the sake of simplicity of treatment.

Pleistocene history of the valley, he found and named the M-Horizon. This is a rubble-bed which occupies an unusual position for such a deposit—the floor of the valley, overlying lacustrine beds. Its own nature and the fact that it is, in turn, overlain by 50 ft. or more of silty deposits, suggested that it was the product of a dry period which caused the desiccation of the previously ponded valley and the conversion of the silts into a land surface.

There is no doubt that, so far, the M-Horizon is the best and most prolific Stone Age horizon in Uganda, for, in most places, the bed is composed of little else but tools—hand-axes, cleavers, cores, core-choppers and flakes. In the course of time this loose breccia of artifacts became hardened by iron solutions and, to-day, may be justly described as an ironstone, or "pan". Thus, though the bed is tantalisingly rich in beautiful tools, it is usually a labour of love and high explosives to extract them.

At the time of our arrival, the best exposure, considered to be M-Horizon *in situ*, was at the big pit excavated by Wayland some years before, of which he published a photograph in 1934.[1] Later on, of course, we dug several sites for ourselves, through the overlying deposits down to the stone-bed, and it was during this work that we realised that there are two parts to the M-Horizon, which we now call A and B, of differing age and tool content. It was, therefore, necessary to qualify the term "M-Horizon" as implying the true, valley *rubble*, in which tools and other stones are quite unrolled, though they might be somewhat weathered.

Thus, at first, we regarded the M-Horizon as a valley rubble, formed under dry climatic conditions, when no water action, that might disturb or roll the implements, was possible in the valley.

Following this definition, it became obvious that the stone-bed at the base of the section in the Big Pit could not be considered as a single, homogeneous deposit of true M-Horizon age. It was only *partly* a rubble and mainly a real gravel, in which tools and stones alike were nearly all water-rolled—some heavily. It then became necessary to decide whether the rolling of this part of the deposit was accomplished during its formation or afterwards, that is, during the post-M-Horizon

[1] *R.R.R.E.M.U.* Plate XLVIII, fig. 2.

124

period of renewed river activity. An examination of the fine-grained beds above the stone layer suggested that there had only been a feeble flow of water during their deposition, which would certainly not have been strong enough to roll the elements of the stone-bed to such a degree as they show.

The final proof that the gravel portion of the Big Pit stone-bed was earlier in formation than the true M-Horizon rubble was provided by the contained artifacts, one group of which (A) was heavily rolled, while the other (B) was little rolled, or fresh. The rolled group is cruder and more primitive than the fresher group, while the latter equates well, both in condition and typology, with the characteristic stage that occurs in the true M-Horizon rubble found at other sites. The presence of this stage, B, at the Big Pit, mixed with heavily rolled tools of the earlier, A stage, immediately suggested that part of the old stone-bed had been water-laid, but had subsequently been exposed gradually to the people of the second stage of culture.

The real significance of this gravel at the Big Pit was not properly understood, however, until we had learnt a great deal more about the geological and climatological events that preceded the M-Horizon. Reference has already been made to the "220 ft." beach, in which, some 15 miles downstream of Nsongezi, Middle Kafuan tools had been found. At this site, the beach-gravel occurs at least 100 ft. above the 100 ft. terrace, while, at Nsongezi, the same beach is barely 50 ft. above the terrace. Furthermore, in and associated with the beach at Nsongezi we found a number of tools similar to those in the gravel at the Big Pit. Consideration of these facts has led to the conclusion that the gravels of the Big Pit and the nearby beach are of about the same age (see p. 49, fig. 3).

It has already been explained that the aggradation of the 100 ft. terrace was the result of Lake Victoria gradually flooding up the valley, during a slow tilt towards the west or south-west. The beach at Nsongezi and elsewhere represents the highest level reached, when the valley was virtually a long, narrow, deep *fjord*, stretching for many miles past Nsongezi. It seems likely that the beach represents a long period of rest at this level.

The Early-Middle Acheulean (Stage A) and Upper Oldowan tools in the Big Pit and beach-gravels suggested the contemporaneity of these two deposits. It is not difficult to suppose that, in the case of the Pit gravels, some of the tools and heavier stones were washed down (for the sides of the valley are steep at this point) to lower levels, where they formed a spread under the water on the silty deposits. It must be admitted that there is hardly any sign, below this layer of gravel, of other gravelly horizons which might, in theory, have been expected to form, from time to time, in the same way. Their absence may, I think, be explained as follows: until the beach period, the water level must have been rising steadily, but, before the greatest effect of the westerly tilt was felt in the valley, the Kagera (and its main affluent, the Orichinga, entering at Nsongezi) was an ordinary if sluggishly flowing stream. Any detritus brought in from the valley sides may well have been removed by the current. Then, when the tilt became more pronounced, rapid accumulation of silt occurred as the ponded river dropped its load further up its course. Finally, the lake invaded the valley, reaching the level of the beach, and any stones or artifacts washed in during the period while it remained at this level spread out over the fine-grained deposits previously laid.[1]

Condition and Typology

Study of all the tools from the Big Pit showed that they fall into three subdivisions according to their state of rolling, A, B1 and B2. The proportions are as follows:

A (very rolled). Core-choppers 83%. Hand-axes 13%. Cleavers 4%.

B1 (less rolled). Core-choppers 52%. Hand-axes 26%. Cleavers 22%.

B2 (little rolled or fresh). Core-choppers 38%. Hand-axes 50%. Cleavers 12%.

Actually, both A and B2 represent only a small proportion of the total number of tools from this site, and Stages B1 and B2 are so close, both in typology and state of preservation, that they may be regarded as belonging to the same stage of culture, which I call B. But A is quite

[1] On this hypothesis, Solomon's postulated connection between Lakes Victoria and Edward (p. 43) would not have taken place until shortly before or during the beach period, but not necessarily even then.

different, first, because so few real Acheulean implements occur amidst a multitude of Oldowan core-choppers, and, secondly, because of the great degree of rolling.

Thus it would appear that Stage A was the industry of Early-Middle Acheulean man, just coming into close contact with Upper Oldowan man. Their tools were washed into the stone-bed at the Big Pit at about the time when the lake stood at the beach (150 ft.) level. Then, as a result of changing climate, the lake began to fall, but, before it had retreated below the stone-bed level, the culture had advanced to Stage B, and tools of this stage were also washed in and slightly rolled. Shortly afterwards the lake fell below this level and the fresh implements of the third group were scattered on and in the previous gravel.[1]

A mile or so north of Nsongezi another site in the M-Horizon produced further evidence of the A–B sequence. Here, again, part of the deposit was a gravel and part a later rubble. The relationship of the two phases was particularly clear at this site (O. 6), as the fresh series was much more numerous here than at the Big Pit. The percentages of tools from both sites are as follows:

Rolled industry; Big Pit stone-bed

Upper Oldowan core-choppers 83%. Hand-axes 13%. Cleavers 4%.

Rolled industry; O. 6

Upper Oldowan core-choppers 70%. Hand-axes 19%. Cleavers 11%.

Fresher tools of Stage B; Big Pit stone-bed

Upper Oldowan type core-choppers 45%. Hand-axes 38%. Cleavers 17%.

Fresher tools of Stage B; O. 6

Upper Oldowan type core-choppers 47%. Hand-axes 26%. Cleavers 27%.

It must be noted that the stages A and B do not correspond *exactly* with the rolled and fresh groups. At both sites, a few of the tools of the later stage fall into the rolled class, since the culture had reached the B stage of development before the waters retreated from the valley floor.

Numerically and typologically, the Big Pit Stage A series bears the

[1] See Appendix A.

closest resemblance to the Jinja series, referred to previously. Here, there were no younger tools in the gravel and the proportions of the A stage were:

Upper Oldowan core-choppers 84%. Hand-axes 16%. Cleavers Nil.

As this was an altogether smaller group, this may account for the absence of cleavers and the slightly larger proportion of core-choppers than occurred at Nsongezi.

The close similarity between the percentages from the two areas, which are nearly 200 miles apart, is strong evidence of the exact morphological state of the Early-Middle Acheulean industry, while the figures also serve to emphasise the significance of the typological differences between this stage and the younger Stage B.

The Stage A tools are of simple type and do not call for much detailed description. The hand-axes tend to be small or medium in size and the majority are of ordinary *biface* type, fairly thick and simply flaked into oval or pear-shaped forms. Even the cleavers are usually made as core-tools, in striking contrast to those of Stage B, but a few are on flakes and foreshadow the great development of parallel-sectioned cleavers of the developed Middle Acheulean. The material used throughout this industry and that of Stage B is a fine-grained, blue-grey, local quartzite, admirably suited to the production of large flake-tools. It has a marked conchoidal fracture and is quite unlike the quartzite employed by the Chellean and Early Acheulean peoples in earlier days. It was also almost exclusively used in the later Levalloisian and Tumbian cultures, in this area.

MIDDLE ACHEULEAN

Apart from its contained Acheulean tools, the M-Horizon of the Kagera 100 ft. terrace is of special importance, since it provides exceptionally clear proof of a marked climatic oscillation. We do not know what the climate was like during the period of silt deposition before this horizon, but the absence of gravels of littoral facies until the beach period strongly suggests the lack of any great rainfall, since heavy precipitation should have resulted in the laying of heavier deposits. Even the valley-gravels, contemporary with the beach (and containing tools of Stage A), cannot be ascribed to the action of a more powerful stream, consequent on increase of rainfall, because the stream had long since ceased to flow, and the valley was occupied by the lake. So, in all probability, the gravels are just the product of lateral erosion and water-sorting during the time when the lake stood at the level of the beach, as explained in the previous section.[1]

The M-Horizon rubble, with its thousands of fresh Stage B artifacts, proves conclusively that the water-level in the valley fell considerably in post-beach times and, further, that no river continued to flow down the valley bottom, following the new, lower base-level, as might have been expected had the fall been due to tilting. Moreover, tilting would have caused marked erosion and there was no sign of this, as far as we could see. Everything points to the valley becoming dry as the result of decreased rainfall, whose first effect was a drop in the level of Lake Victoria.

Although the industry of the M-Horizon rubble is more advanced than that of the beach period, it is not essentially different, so it is unlikely that a long period elapsed between them. Consequently it would appear that, once it set in, the desiccation of the valley was comparatively rapid.

Technique and Typology

The Stage B industry includes hand-axes, cleavers, core-choppers, flakes and rare scrapers and large cores.

The hand-axes and cleavers of this stage are of very characteristic type. For the first time, in Uganda, large flakes were used almost exclusively in the manufacture of these tools. The primary flakes were either struck directly off outcrops of quartzite, or, more probably, off large cores, and then, by simple retouching, converted into beautiful hand-axes and cleavers. In many of the implements, the bulb of percussion—usually at one side—was flaked away, and similar flaking on the opposite edge reduced the original length of the flake and made it shapely. This method almost invariably produced a parallelogram section in the finished artifact and is a modified form of the Victoria West technique, and one that was widely employed from one end of Africa to the other. Its use is, of course, particularly evident in the

[1] See, however, Appendix A on this explanation.

cleavers, for which only the minimum amount of retouching was necessary, to reduce the edges of the primary flake to suitable proportions for grasping or hafting.

There is no trace of the true Victoria West technique of large, *prepared* cores, but a few, very rare instances of another method occur in the M-Horizon. With this, one edge of the core was trimmed for use as a striking platform by detaching a number of short, steep flakes, so that the resulting large flake also had a "facetted" platform. Also, occasionally, a few of these scars tended to "run" across the surface of the core, and, if a large flake were later detached at right angles, its face received a pseudo-Levalloisian appearance of primary preparation. I am quite sure that only the first of these two actions was intentional and the second purely fortuitous and, in any case, both are so rare as to have no special technical significance. They certainly do not imply the birth of the Levalloisian technique, as is claimed tentatively for the Victoria West. For the purpose of describing our simple core and flake technique for the production of this type of hand-axes and cleavers, and for the tools themselves, however, the term Victoria West is, I think, appropriate. Hand-axes and cleavers flaked or retouched on both sides, in the usual, western European manner, are here referred to as *biface* tools.

The following percentages are based on the largest of our M-Horizon collections, and show the relative proportions of the Victoria West and *biface* types of tools:

Hand-axes

Victoria West, parallelogram type	57%
Biface type	43%

Cleavers

Victoria West, parallelogram type	78%
Biface type	22%

Thus, the proportion of tools of *biface* type to those of parallelogram form is about 33 to 67%.

Neither type of hand-axe calls for much individual description, as nearly all the tools are ordinary pointed ovates or *limandes* and except for a few specimens of the *biface* type, one of which is 13 in. long, are quite unremarkable, but the Middle Acheulean cleavers exhibit a variety of shapes that are worth recording. They are of medium size, with three types of blade, all of whose butts are variously squared, rounded or pointed, as follows:

    1. Square-bladed: the normal type of cleaving edge.

    2. Guillotine-bladed: the edge slants steeply, forming one acute and one obtuse angle with the sides of the tool.

    3. Narrow chisel-bladed: the edge is much narrower than the width of the rest of the tool, which thus resembles an unfinished hand-axe. These blades are sometimes only $1\frac{1}{2}$ in. wide. This type of tool is usually made by the *biface* technique, though on a flake.

True scrapers are rare in this industry and, in view of the degree of skill shown in the manufacture of hand-axes and cleavers, it seems reasonable to suppose that scrapers would have been numerous also, had the people found any great need for them. As it is, probably un-worked flakes and other sharp stones were used more or less haphazard for cutting and scraping purposes.

There is a large class of core-choppers and the like, which form about 47% of the total worked objects, at the largest of our sites. There seems to be no doubt that this group is descended from the true Upper Oldowan core-choppers of Stage A, but their variety in Stage B is considerable, whereas the Stage A choppers all belonged either to the spherical or cuboid types. Nearly all the Stage B core-choppers are both smaller and more finely and extensively chipped than those of Stage A, and show more ability and purpose, whereas the earlier implements tend to be rather haphazard in workmanship.

Only a few of the big cores were found, from which the large, primary flakes were obtained, but this is not surprising, considering the size of core necessary for obtaining flakes big enough for conversion into hand-axes or cleavers; it is more than likely that the bulk of the primary flaking was done at the site of the quartzite outcrop, where the material was obtained. Many of the originally large cores were turned into core-choppers as soon as they became too small to produce flakes large enough to make hand-axes or cleavers. As we found practically no small, retouched flakes, we cannot regard the reduction in size of the large cores as a natural process due to continued use in the production of utilisable flakes of all sizes, so it seems to have been a deliberate method for obtaining core-choppers. The largest true core that we found measured approximately 29 × 26 cm. (about $11\frac{1}{2} \times 9\frac{1}{4}$ in.).

Personal experiment and the examination of platforms and bulbs of percussion showed that the best—indeed, almost the only—method of obtaining large flakes in quartzite of this type was to use the "Clacton" technique of swinging the core down on to a fixed anvil-stone. Direct percussion with a hammer-stone was almost useless, the usual result being only a painful wrist and skinned knuckles.

Considering the beauty of many of the Middle Acheulean tools and the comparative flatness of the retouching, I was a little surprised at the opinion of M. Coutier, when he examined our material in Paris, that nearly all of it was stone-flaked, although this bore out what had been suggested by experiment in Uganda. I had found that passably flat flaking could be produced in retouching hand-axes by subjecting the tool to the Clacton method of percussion on an anvil. With a little practice, sizable, fairly flat flakes could be detached when roughing out the shape of the tool, and then further flakes could be removed along the ridges formed by the intersection of earlier scars. This method reduced the thickness of the tool, giving it a flattish appearance, with fairly straight edges. With implements made by the Victoria West method, with simple edge-flaking for the retouch, only direct percussion was required, and the resultant flake-scars were short and steep.

These facts show that the use of wood technique is by no means such a valuable criterion for recognising the Acheulean age of hand-axe industries as some workers have claimed.

It is very necessary to take into account the quality of the raw material when studying the incidence of wood technique in a hand-axe industry, for its rarity or absence is not necessarily a sign of the primitiveness of the industry; that must be judged by the quality of the tools generally and by comparison of their types with those in nearby areas. Thus, though the Early Acheulean (Stage 1) of Oldoway Bed III *may* contain a higher proportion of wood-flaked tools than the M-Horizon industry, that fact does not make the latter as early, or earlier. The important factor is whether or not the degree of skill generally exhibited by the M-Horizon stage matches that of the Oldoway Bed III stage and, especially, if the tool-forms are the same.

The Acheulean

The M-Horizon Acheulean, as we realised after close comparison, is definitely later than the Bed III stage, and, in the variety of its tool-types—especially those produced by the Victoria West technique—is obviously most closely comparable with Oldoway Acheulean Stage 4, from Bed IV. The only marked difference is the abundance of inherited Oldowan types in the M-Horizon industry.

UPPER ACHEULEAN

Upper Acheulean industries seem to be rare in Uganda, and unknown to us in stratified deposits. The latter fact is of more than local significance, for, even in Kenya, the Upper Acheulean does not occur in bedded deposits, but in rubbles, while at Oldoway it is found at or near the top of Bed IV, whose deposition was very largely accelerated by the inclusion of wind-borne volcanic ash, etc., in an area especially favourable for subaqueous deposition.

In Uganda we know of no deposits of the same age as the upper part of Oldoway Bed IV, and, even in such comparatively favoured regions as the Kagera valley, the latest Lower Palaeolithic silts are those underlying the M-Horizon, in the Kagera 100 ft. terrace. The post-M-Horizon deposits appear to be entirely of Upper Palaeolithic age.

There is only one stage of Acheulean in the true M-Horizon rubble, and we know that the geological break between this deposit and those that follow it must have been of considerable length.[1] Where is the Upper Acheulean? There is no sign of any such industry following the Middle Acheulean in the Kagera valley, and I attribute this to the intensity of the drought that caused the desiccation of the valley during M-Horizon times. On the other hand, there is no evidence that Lake Victoria dried up completely during this period, in fact, such an event is most unlikely, so that it is to this vicinity that we should expect Upper-Middle Acheulean man to have migrated when increasing desiccation in the river valleys further inland became intense enough to affect meat and water supplies. So much was suggested by theory, and it seemed to be borne out in practice when we discovered two widely separated Upper Acheulean sites, both close to the lake. Even these were not in lake silts, but in rubbles.

[1] See Appendix A for further discussion of this point.

133

PLATE IX

Acheulean Culture

Two Early Acheulean hand-axes belonging to the oldest series in the Younger Rubble, Kagera valley, near Nsongezi; quartzite, weathered. Note the pronounced base-keel on the lower implement. Both these tools are repeated on p. 139 (1 and 2).

Approx. ½ scale.

1

2

FIGURE 17

Acheulean Culture

No. 1: A base-keeled Early Acheulean hand-axe belonging to the oldest series of tools in the Kagera valley Younger Rubble. Unworked on the two lower faces; quartzite, weathered.

No. 2: A bi-convex Early Acheulean hand-axe from the same provenance; quartzite, weathered.

No. 3: A flat-based Early Acheulean hand-axe from the same provenance; quartzite, weathered.

$\frac{1}{2}$ scale.

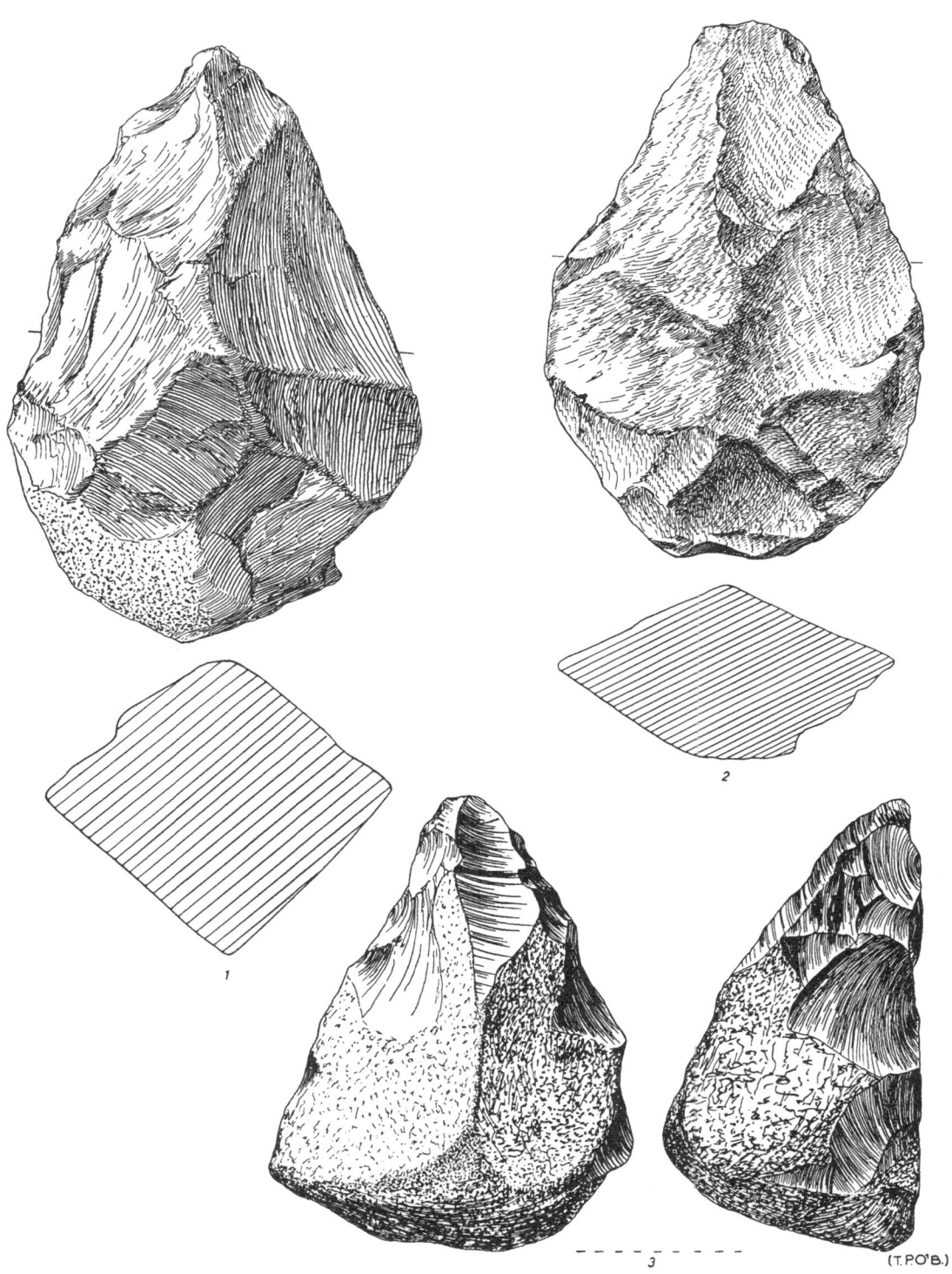

(T.P.O'B.)

FIGURE 18

Acheulean Culture

Three further examples of Early Acheulean tools from the Kagera valley Younger Rubble; quartzite, weathered. No. 1 is a base-keeled tool with the two lower faces unworked. Note the two cones of percussion in No. 3 which is a hand-axe made on a flake. The underside is base-keeled and unworked; all quartzite, weathered.

$\frac{1}{2}$ scale.

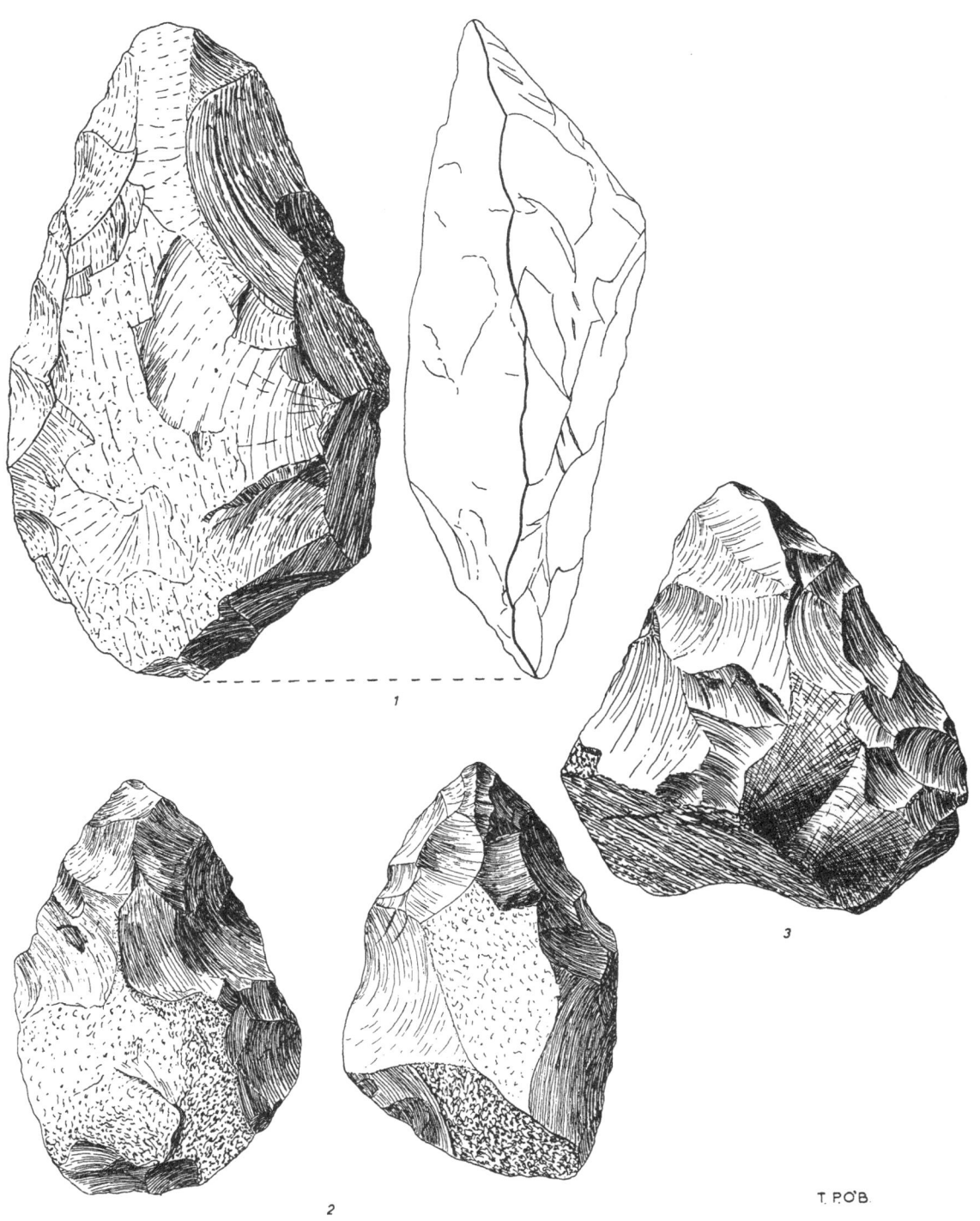

1

2

3

T. P. O'B.

141

PLATE X

Acheulean Culture

Fig. 1. Two views of an Early-Middle Acheulean cleaver from the Phase A gravel in the M-Horizon, Nsongezi (Kagera 100 ft. ± terrace); quartzite, rolled.

Fig. 2. Two views of a hand-axe made on a small boulder, from the same locus as (1); quartzite, rolled.

Approx. ½ scale.

1

2

143

PLATE XI

Acheulean Culture

Fig. 1. Two views of a large Early-Middle Acheulean hand-axe from gravels (beach?) between Nsongezi and Kikagati; quartzite, rolled. Repeated on p. 147 (1).

Figs. 2 and 3. Tools found in the Early-Middle Acheulean beach gravel at Mile 14, Nsongezi-Mbarara Road; quartzite, rolled. No. 2 repeated on p. 149 (6).

Approx. $\frac{1}{2}$ scale.

1

2   3

FIGURE 19

Acheulean Culture

Rolled Early-Middle Acheulean hand-axes.

No. 1: From the Phase A beach gravels between Nsongezi and Kikagati; quartzite.

Nos. 2, 3, 4 and 5: Unusually small examples obtained by Mr E. J. Wayland and loaned by him for figuring. 2, 4 and 5 came from the M-Horizon, Kagera 100 ft. terrace, at Nsongezi and, judging from their rolled condition from the Phase A part of it (5 is more rolled than the drawing suggests). 3 comes from the Early-Middle Acheulean beach at Mile 14, Nsongezi-Mbarara Road; all quartzite.

$\frac{1}{2}$ scale.

146

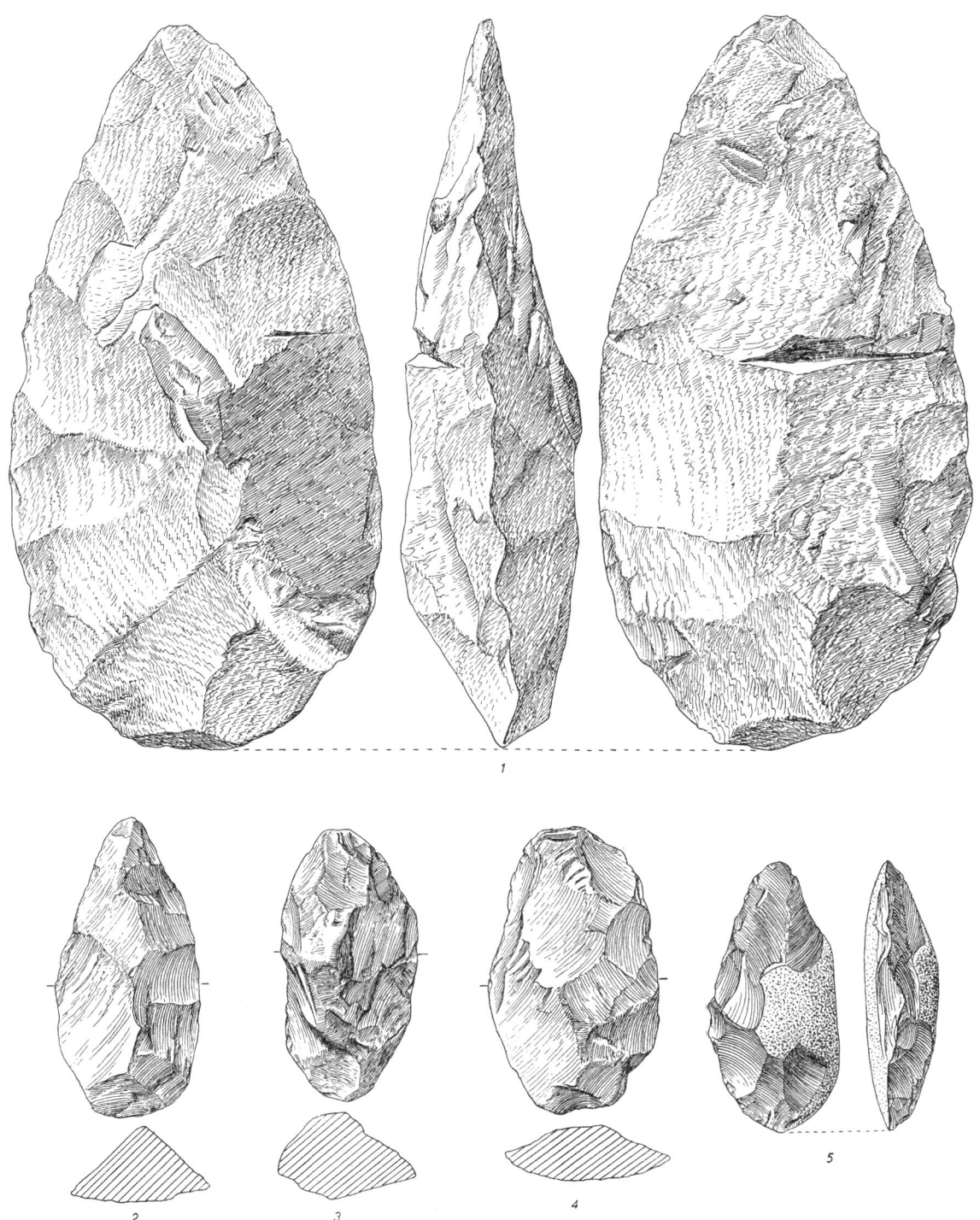

1

2   3   4   5

10-2

FIGURE 20

Acheulean Culture

Nos. 1, 2, 3 and 5: Upper Oldowan core-choppers in association with the Early-Middle Acheulean industry. From the Phase A gravel in the M-Horizon, Kagera 100 ft. terrace, at Nsongezi.

No. 4: An Early-Middle Acheulean cleaver from the same provenance.

No. 6: A discoid artifact from the Early-Middle Acheulean beach at Mile 14, Nsongezi-Mbarara Road; all in quartzite, strongly rolled.

½ scale.

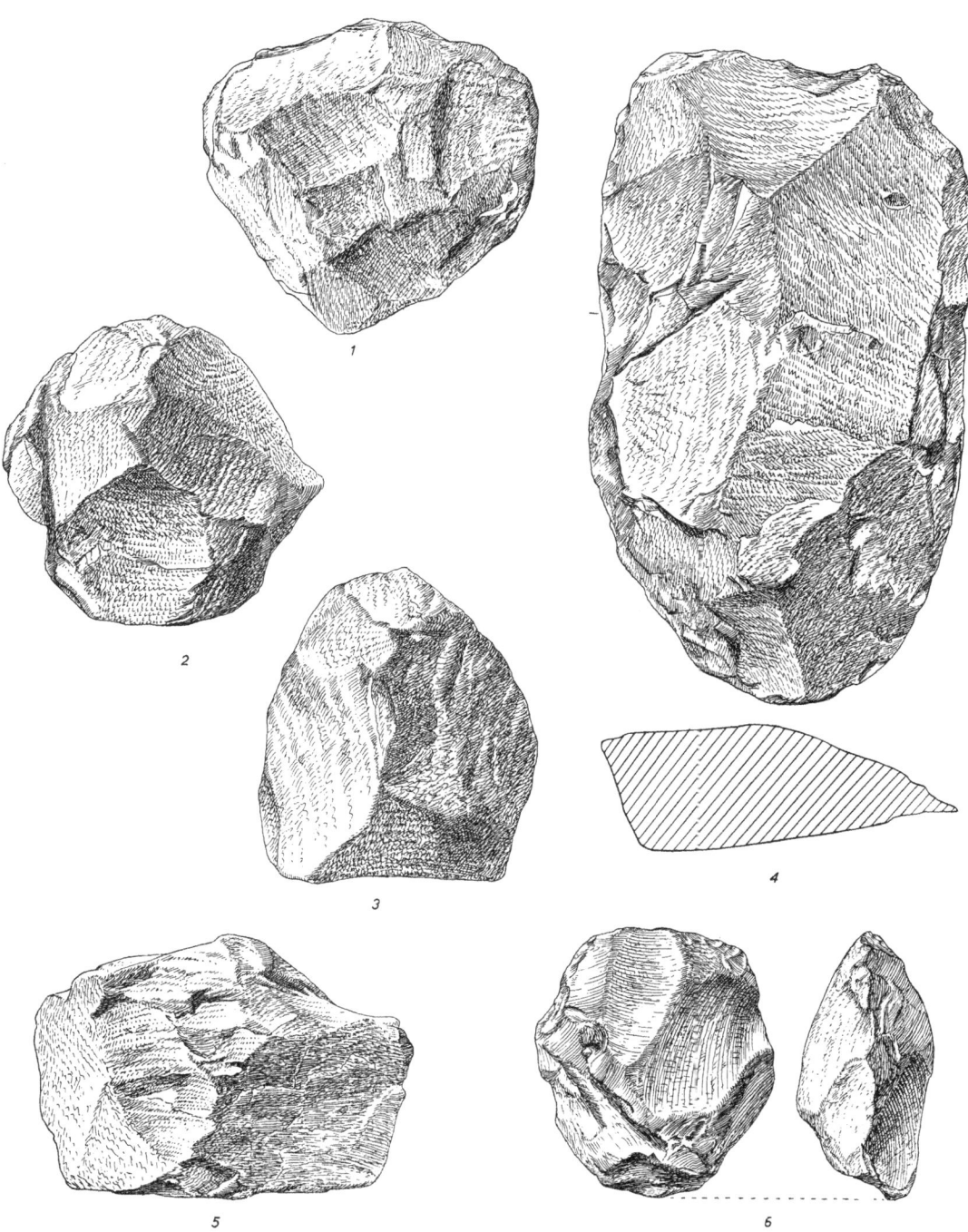

PLATE XII

Acheulean Culture

Two views of a large, flat Middle Acheulean *biface* from the Phase B rubble in the M-Horizon, near Nsongezi; quartzite, fresh. Repeated on p. 155.

Approx. $\frac{1}{2}$ scale.

151

PLATE XIII

Acheulean Culture

Fig. 1. Two views of a Middle Acheulean hand-axe from the Phase B rubble in the M-Horizon, near Nsongezi; quartzite, fresh.

Fig. 2. Two views of a Middle Acheulean cleaver made on a side-blow flake, found in the Phase B rubble in the M-Horizon, near Nsongezi; quartzite, fresh. Repeated on p. 159 (1).

Approx. ½ scale.

1

2

153

FIGURE 21

Acheulean Culture

Three aspects of a large Middle Acheulean *biface* from the Phase B rubble in the M-Horizon, Kagera 100 ft. terrace, near Nsongezi. The general flatness of the flaking suggests the use of a wood technique but I am assured by the French expert, M. Coutier, that the implement was stone flaked on a large flake; quartzite, fresh.

$\frac{1}{2}$ scale.

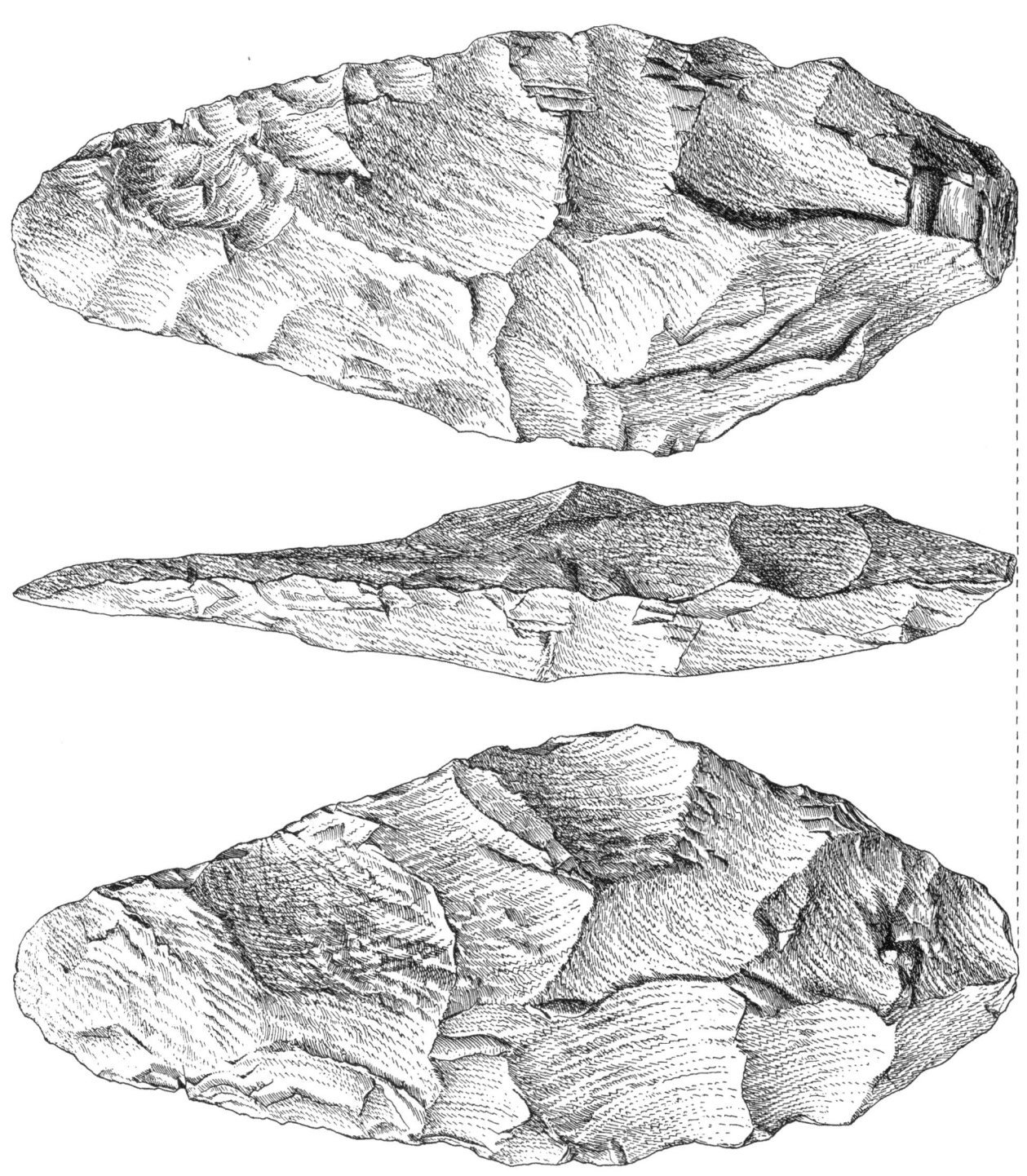

FIGURE 22

Acheulean Culture

Three Middle Acheulean cleavers from the Phase B rubble in the M-Horizon, Kagera 100 ft. terrace, near Nsongezi; quartzite, fresh.

½ scale.

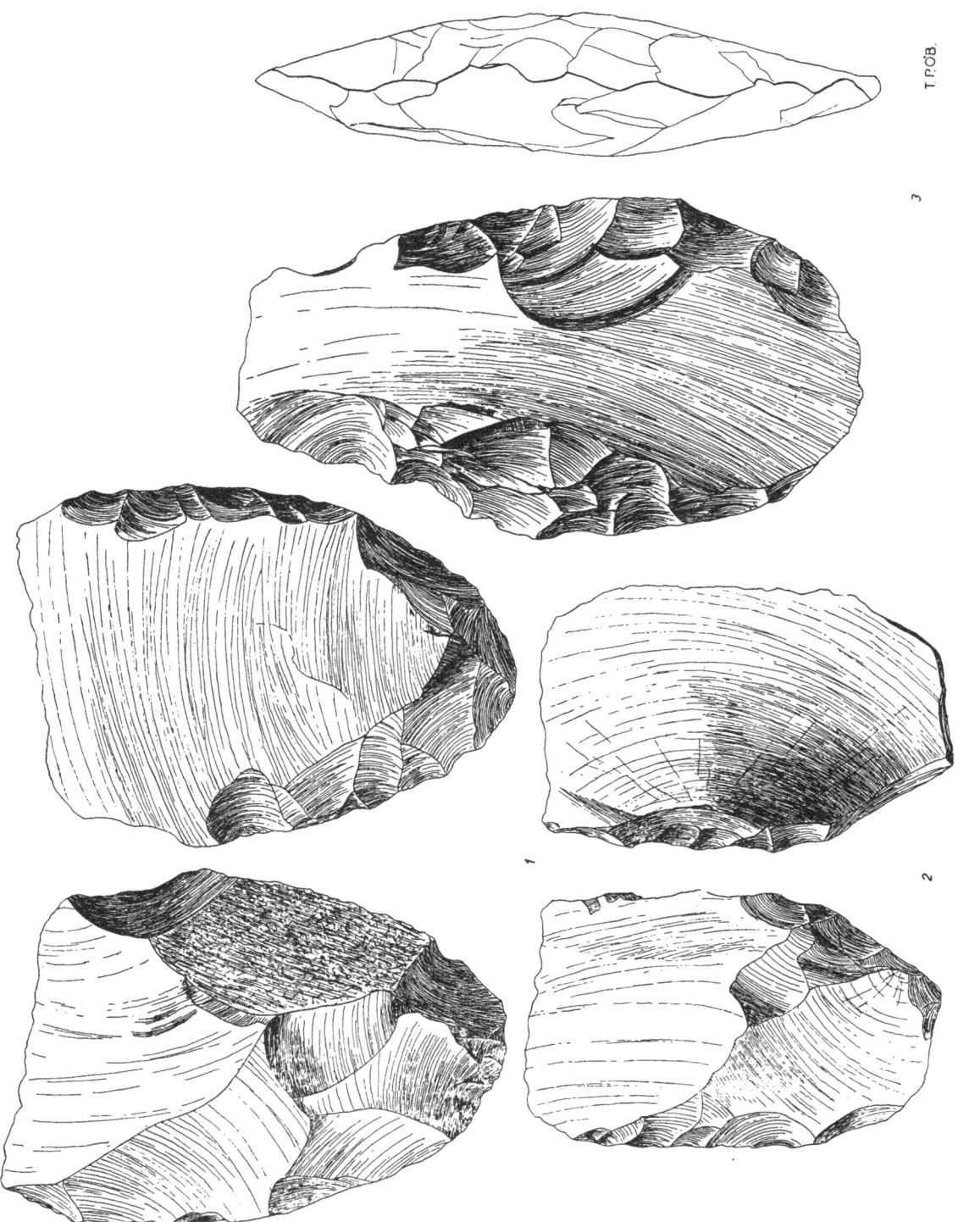

157

FIGURE 23

Acheulean Culture

No. 1: Middle Acheulean cleaver of "guillotine" type from the Phase B rubble in the M-Horizon, Kagera 100 ft. terrace, near Nsongezi; quartzite, fresh.

Nos. 2 and 4: Middle Acheulean hand-axes from the same provenance; quartzite, fresh.

No. 3: A Middle Acheulean cutting tool from the same provenance; quartzite, fresh.

$\frac{1}{2}$ scale.

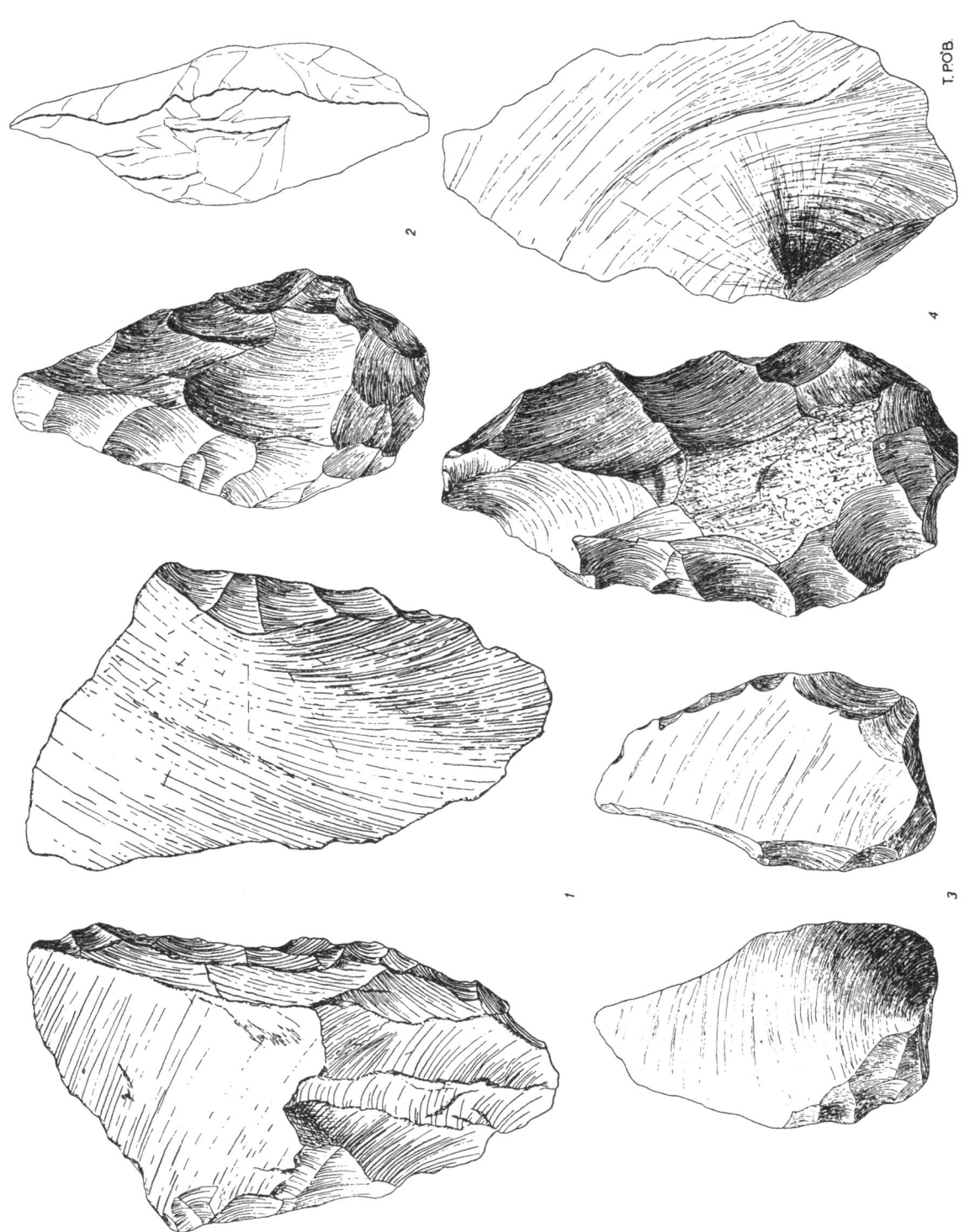

FIGURE 24

Acheulean Culture

Nos. 1, 2, 3 and 4: Middle Acheulean core-choppers descended from the Upper Oldowan. From the Phase B rubble in the M-Horizon, Kagera 100 ft. terrace, near Nsongezi; quartzite, fresh.

No. 5: A small core from the same provenance; quartzite, fresh.

No. 6: A discoid artifact from the same provenance; quartzite, fresh.

No. 7: An end-scraper from the same provenance; quartzite, fresh.

No. 8: A small point made on a flake from the same provenance; quartzite, fresh.

$\frac{1}{2}$ scale.

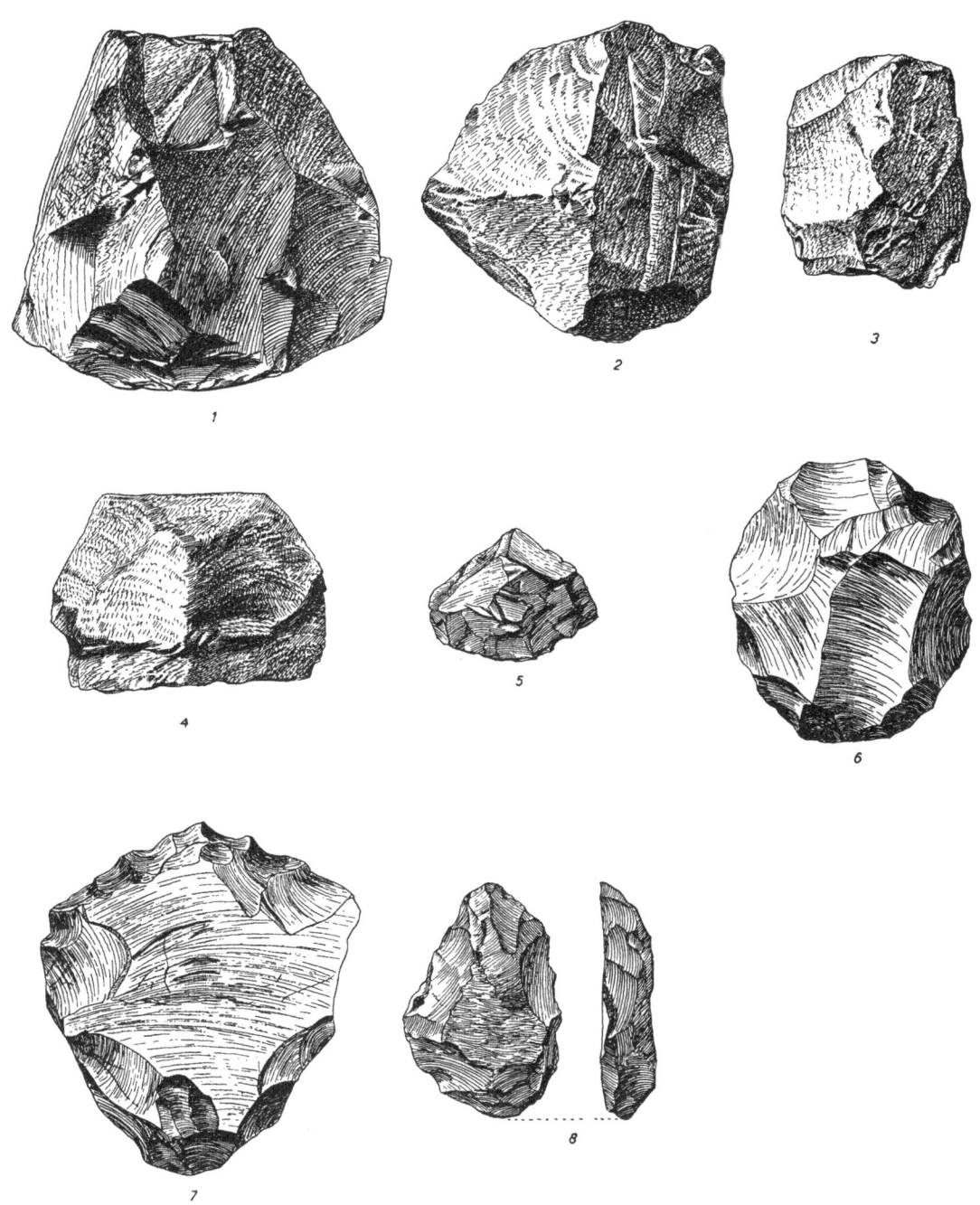

PLATE XIV

Acheulean Culture

Figs. 1 and 2. Two Upper Acheulean hand-axes from a rubble on the Sango Hills, Lake Victoria, near the mouth of the Kagera river; quartzite, somewhat weathered. Repeated on p. 165.

Figs. 3, 4, 5 and 6. Upper Acheulean hand-axes from a rubble at the Railway borrow pit, near Bugungu, Lake Victoria; quartzite, weathered. Repeated on p. 167.

Approx. $\frac{2}{3}$ scale.

FIGURE 25

Acheulean Culture

Two Upper Acheulean hand-axes from a rubble on one of the Sango Hills, near Lake Victoria. Their state of preservation shows that they belong to the second oldest series in this locality.

$\frac{2}{3}$ scale.

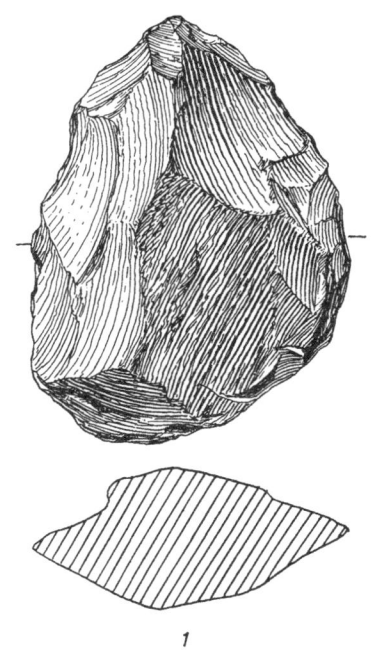

1

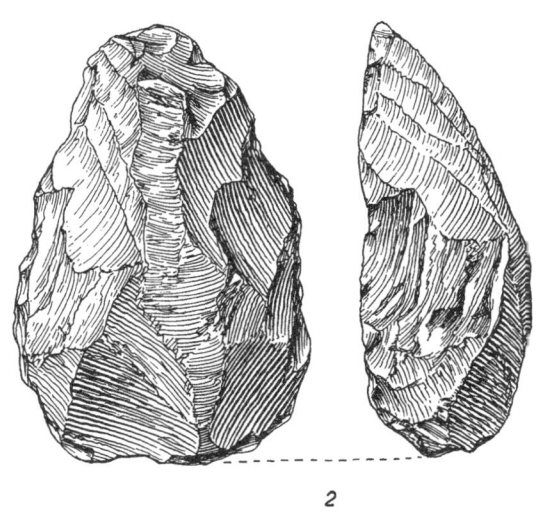

2

165

FIGURE 26

Acheulean Culture

Four Upper Acheulean hand-axes from a rubble at the Railway borrow pit site near Bugungu, near Lake Victoria. Their state of preservation shows that they belong to the second oldest series present in this rubble and antedate the local Proto-Tumbian and Levalloisian industries.

$\frac{2}{3}$ scale.

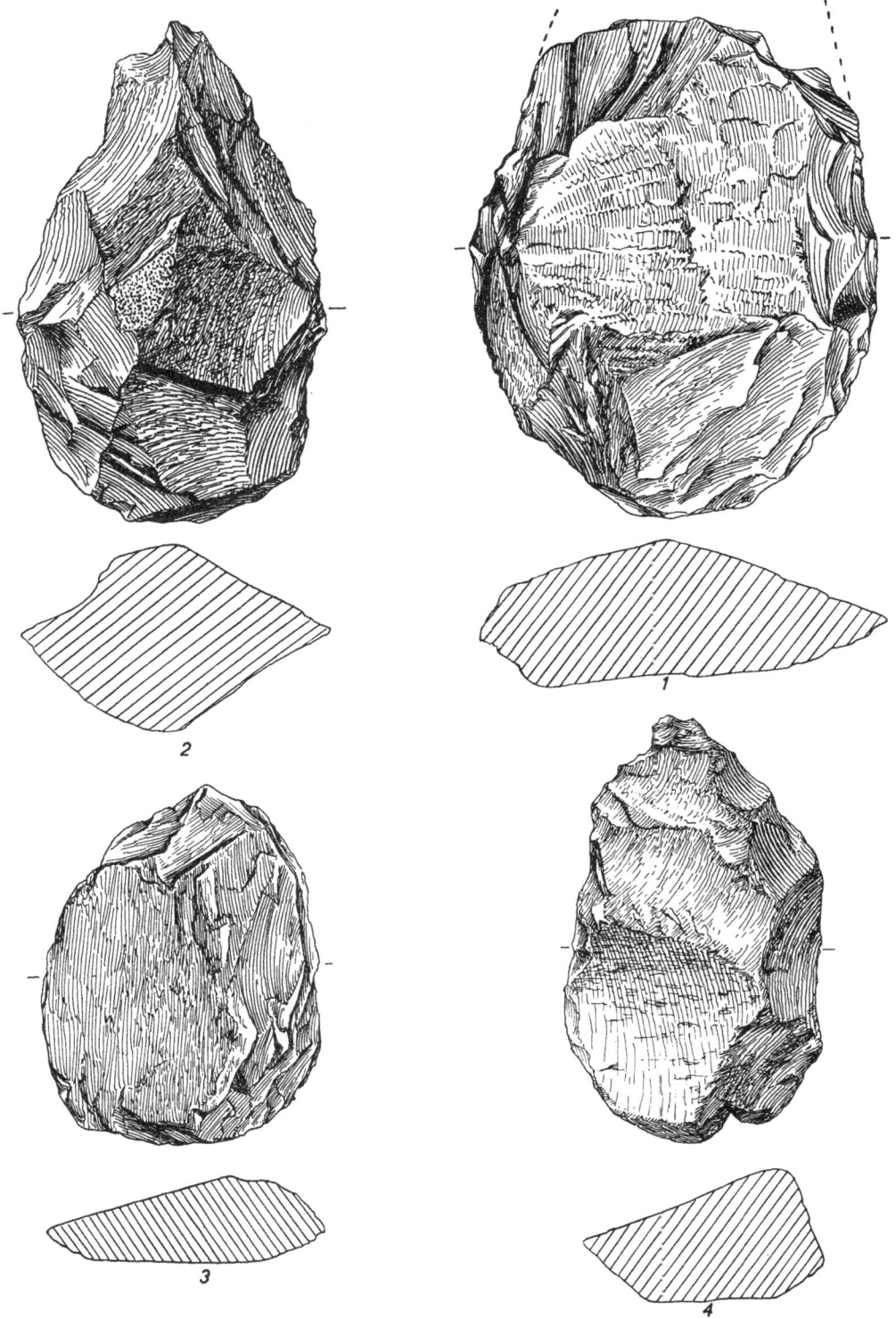

The most numerous (though still scanty) examples of Upper Acheulean tools occur as a single, easily recognisable group among a number of cultural stages—mainly Levalloisian—from a rubble near Jinja[1] (see p. 67, fig. 3). At this site the only way of dividing the very mixed assortment of industries was by typology and state of preservation, and this method showed that the Levalloisian cores and flakes were all younger than the small Upper Acheulean hand-axes (there were no cleavers), which have a characteristic, deep, reddish stain, derived from the underlying lateritic rubble. Also, owing to the original poor quality of the local quartzite, the hand-axes are not very well made, and are now so rotted as to require very careful handling, whereas even the oldest Levalloisian tools are much less deeply stained, and still in a good state of preservation.

The second place where the Upper Acheulean is found is the rubble on the Sango Hills, at the mouth of the Kagera. Here, again, several industries occurred and the tools could only be classed on the basis of their state of preservation, but this was confirmed by typology. In this assemblage, the small hand-axes (rather better made than the Jinja group) constitute almost the oldest series—a fact that might have surprised us considerably, in view of the comparative freshness and extreme crudity of the bulk of the "Sangoan" material which accompanied it. But, as I have explained in Chapter IX, although much of that crude material had previously been regarded as of Lower Palaeolithic date, we now know that it belongs to a Proto-Tumbian industry which was of Upper Palaeolithic age.

At present, all that we can say about the dating of the Upper Acheulean is that it appears to belong to some part of the dry period which began, locally, in Middle Acheulean times, and antedated the earliest Levalloisian.

Technique and Typology

There is little to say under this heading, for the material is very scanty. What there is, may be simply described as an industry of small hand-axes of oval or cordiform shape, made on flakes.

---

[1] Regarded as M-Horizon by Wayland: see *R.R.R.E.M.U.* Plate XLIX, fig. 2.

CHAPTER X

The Levalloisian and Still Bay

THE origin of the African branch of the Levalloisian culture is, at present, a mystery, which the rival claims of Europe and South Africa have done nothing to solve. If the dating of the East African branch in relation to the Acheulean means anything, the Levalloisian made its first appearance there later than it did in South Africa, where it was the contemporary of several stages of Acheulean, though these may themselves have existed contemporaneously. If the Levalloisian went to South Africa from the north, it would seem to have followed some other route than that through the eastern half of the continent, but, if the culture originated in South Africa, its comparative lateness in arriving in East Africa is understandable, though, in that case, it should have appeared still later in Europe.

The South African claim for the origin of this culture is based on the appearance of the prepared core and flake technique of the Victoria West Variation in Upper Stellenbosch, that is, Upper Acheulean times. In fact, this industry is sometimes referred to as proto-Levalloisian. Anyone who has examined Victoria West cores must have been struck by their resemblance to those of the Levalloisian in their primary preparation, although the large flakes detached from them were converted into ordinary African Acheulean hand-axes and cleavers. On the other hand, it is probable that most African Acheulean industries used some kind of core and flake method, since the majority of the hand-axes and almost all the cleavers were made on flakes. It is not improbable that simple preparation of the core for the easier detachment of flakes was an easy and early discovery. In Uganda it is even apparent in the Middle Acheulean, though very rarely used.

If it can be proved that industries of the Victoria West type gradually led to a pure Levalloisian culture, by the dropping out of hand-axes and cleavers and the intensification of Levalloisian characters, then it may be necessary to reconsider the European claim to the culture's earliest appearance. So far as it is possible to compare the European and South

169

African cultures and periods from the standpoint of their relative ages, however, it seems to me that the Levalloisian of western Europe is much older even than the Victoria West variation, and most decidedly older than the first, *true* South African Levalloisian.

EAST AFRICA

Despite recent statements,[1] which are not supplemented by any kind of proof, regarding the earliest East African Levalloisian, there seems to be no earlier stage than that represented in Kenya by the Nanyukian. Even at the type station, however, the association of Levalloisian with Upper Acheulean tools occurred in a rubble which cannot be precisely dated with reference to the Rift Valley deposits. The claim that the Nanyukian was also to be found under Kamasian sediments, on the edge of the Kinangop Rift escarpment, is not one that I can support, since personal examination of the sites showed that the tools occurred in a rubbly deposit below post-Kamasian wash-earths, which are dated by Still Bay tools. Moreover, it has yet to be proved that the Nanyukian Acheulean tools are of East African Lower Palaeolithic age, for there is a possibility that they are late survivals into the lower part of the Upper Palaeolithic.

In Uganda, the Levalloisian culture is the most pervasive of any, and occurs in almost every corner of the Protectorate. Poor quality or paucity of material in many places was, apparently, no deterrent to the widespread colonisation of this culture. Yet, in spite of this robust, pioneering spirit, which might have carried it to great heights of cultural achievement, the Uganda Levalloisian remained a monotonously conservative and unspecialised complex throughout nearly the whole of its long history. Only at the end did some part of it break away from a monotonous tradition of cores, flakes and little else, to flourish, for a time, as the Still Bay.

Kagera Valley

The most abundant stratigraphically dated Levalloisian in Uganda is that which we found in the post-M-Horizon, riverine and lacustrine beds of the Kagera 100 ft. $\pm$ terrace.

[1] Leakey, *Stone Age Africa*, pp. 47 *et seq.*

The Levalloisian and Still Bay

Originally, we regarded these beds as of Upper Kamasian age, partly because the magnitude of the break between them and the M-Horizon was not then appreciated, and partly because, in them, we found tools which we then thought were Upper Acheulean. This dating meant that we had, in Uganda, a very prolific and well-developed Levalloisian, much more advanced than the Kenya Nanyukian and apparently even earlier.

We now know that this idea was almost certainly wrong, although absolute proof is not forthcoming yet. The, apparently, Acheulean hand-axes are really part of a distinct culture—an early form of Tumbian—which has every appearance of having descended indirectly *from* the Acheulean, and this and the advanced typology of the Levalloisian (which remained a distinct, though contemporary, group) indicates with virtual certainty that this part of the 100 ft. terrace deposits is of Gamblian, that is, East African Upper Palaeolithic age.[1]

This view necessarily involves a long break between the Middle Acheulean M-Horizon and the succeeding Levalloisian beds—a gap in the geological history of the valley which must have been of climatic significance, since there is no sign of intervening "wet" deposits. It must mean that the dry period that drove Acheulean man out of the valley was of considerable length and, presumably, severity. Fortunately, the Kenya evidence of a similar break, between the Upper Kamasian and the Lower Gamblian, affords very strong support to this theory. There, however, the break either began later than it did in Uganda, or else only the maximum peak of aridity was effective in causing desiccation and related phenomena. This might have been caused by the geographical differences in height and topography between the two countries. However that may be, the whole of East Africa seems to have been in the grip of a severe dry climate by the time of the Upper Acheulean culture stage. The results varied in different places, of course, depending on the mean rainfall variations between different regions, and this may explain why, while Uganda rivers such as the Kagera dwindled to a mere trickle, or dried up altogether, parts, at least, of Kenya and Tanganyika were still habitable and possessed good water supplies until very much later.

[1] In the sense of time, not culture, of course.

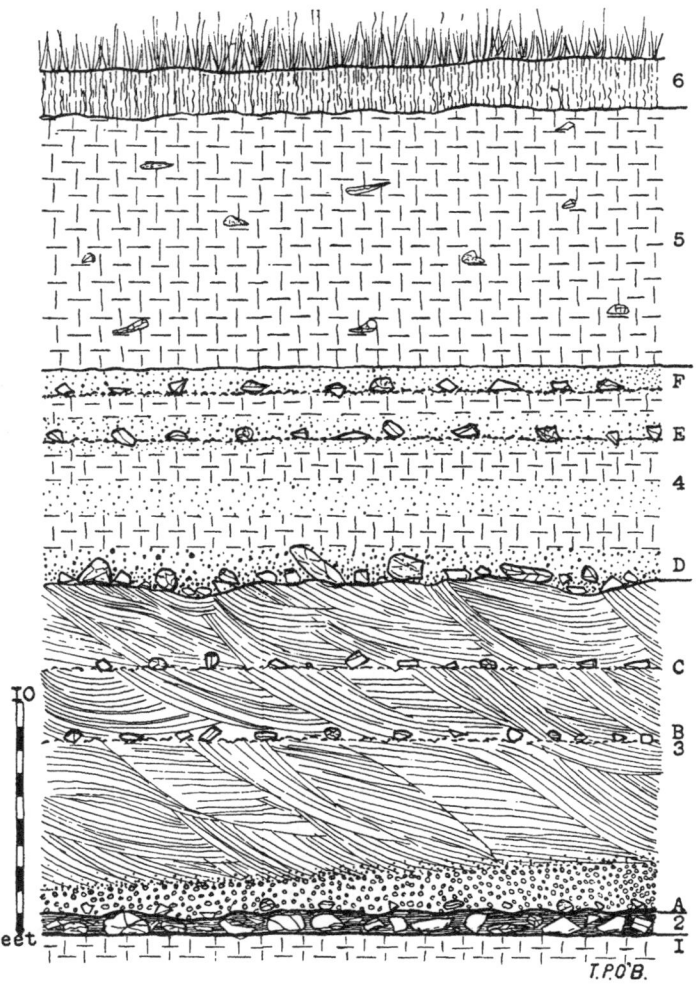

FIG. 27. Section showing upper half of Kagera 100 ft. terrace series. Pit in Orichinga Valley 2 miles north of Nsongezi.

6. Fine, light brown surface soil; 2 ft.

5. Pale greenish or greyish-brown clay; 11 ft. Tumbian and Levalloisian tools throughout.

4. Bed of alternate layers of fine white sand or clay; 9 ft. N-Horizon Rubble at base. Unconformable with 3.

3. Current-bedded yellow sands, shingly at base; 14 ft. Unconformable with 2.

2. M-Horizon Rubble, ferruginised; 1 ft.

1. Clay, top of lower half of Kagera 100 ft. terrace series.

A, B, C, D, E and F are thin horizons (probably short-lived land-surfaces) producing Levalloisian implements. In D, Levalloisian is mixed with proto-Tumbian (N-Horizon).

172

The Levalloisian and Still Bay

The return of normal climate to East Africa was marked everywhere by the laying of a new series of riverine and lacustrine beds. In the Kenya basins the Gamblian sediments began to form, while in the Victoria basin the increasing rainfall caused the rivers to flow and the lake to rise again. Before it rose sufficiently to flood the Kagera valley as far as Nsongezi, however, a considerable thickness of false-bedded sands had been deposited by the sluggishly flowing stream on top of the old, hardened M-Horizon.

Then a minor dry oscillation occurred, resulting in the formation of a small rubble—the N-Horizon—and this was followed by the deposition of a bed of fine white sands, containing bands of clay. Ultimately, the maximum rise of the waters, caused by the lake flooding up the valley again, led to the laying of a thick homogeneous bed of clay which is the top and last bed of the 100 ft. terrace succession. The lake, however, never reached as high up the valley as it did in the days of the 220 ft. ± Early-Middle Acheulean beach and the topmost clays are more in the nature of swamp deposits than deep-water sediments.

A number of Levalloisian horizons were found *in situ* at various levels throughout these deposits, and a very late stage occurred in the thick alluvial filling of a main tributary valley, which was cut through the 100 ft. terrace after the last earth movement. This last stage showed a trace of Still Bay.

Post-M-Horizon False-bedded Sands

The first stratigraphical Levalloisian stage occurred near the base of these sands, at a depth of about 36 ft. from the top of the terrace. It was found at two different sites, in the same position. The proportion of facetted flakes was 15%.

The second horizon was a thin one, found at one site, at a depth of 29 ft., and contained 25% of facetted flakes.

The third horizon occurred at the same site as the second, at a depth of 26 ft. There were 28% of facetted flakes.

N-Horizon and N-Horizon Bed

The fourth Levalloisian level was the N-Horizon rubble, 22 ft. from the surface, which terminated the false-bedded sands and was found at a number of sites. The Levalloisian tools formed a distinct group from the Proto-Tumbian implements which accompanied them in this deposit, being made of the usual blue quartzite, while the latter were made of a pale grey variety. The proportion of facetted flakes was 35%.

The fifth Levalloisian horizon was a thin layer in the N-Horizon Bed of white sand and clay, about 6 ft. above the N-Rubble, and 16 ft. from the surface. Facetted flakes here formed 42%.

The sixth horizon also occurred in the N-Horizon Bed of sand and clay, 14 ft. from the surface. Most of this group was in white quartz, instead of the usual blue quartzite and the facetted flakes totalled 50%. There was no Proto-Tumbian in these last two levels.

Uppermost Clays

In the stiff homogeneous clays above the N series, there are no definite tool horizons, but thousands of implements lie scattered through it at all levels. The valley bottom was evidently a marsh, at this time, over which man was occasionally able to roam. Levalloisian cores and flakes, as well as quantities of beautiful Tumbian implements, were abundant in the numerous washouts where this clay has been cut through down to the top of the N series. The top of these N sands almost invariably forms the floor of these erosional features. As the tools were all mixed indiscriminately together, it is not possible to give any figures for the Levalloisian of this level, since many of the plain flakes must belong to the Tumbian industry, but, judging by the quantity of really good, typical Levalloisian cores and flakes with facetted platforms, the culture was more advanced than at any time previously.

Technique and Typology

Apart from the gradually increasing proportion of flakes with prepared striking platforms, the Levalloisian from the Kagera 100 ft. terrace is a monotonously homogeneous culture. It is certainly not possible to treat all the horizontal stages as cultural divisions, but the clear geological differences between the various deposits affords an easy, if somewhat arbitrary, means of dividing the culture into three main stages, as follows:

The Levalloisian and Still Bay

Post-M false-bedded sands, 17 ft. Levels 1, 2, 3, Lower Levallois.

N stone-bed, clays and sands, 12 ft. Levels 4, 5, 6, Middle Levallois I.

Top clay, with Tumbian, 17 ft. Level 7, Middle Levallois II.

There is little to be said about the typology of the Kagera valley Levalloisian. The only real implements were scrapers and points, neither of which were common, while chunky core-choppers and hammer-stones were fairly numerous. The bulk of the culture consisted of flakes with plain or facetted striking platforms and cores. The majority of the latter are of discoid and not tortoise type, and many of them are very small. Such "discs" are present in nearly all African Levalloisian industries, and several suggestions as to their supposed use have been put forward, for instance, that they were sling-stones. I myself think that such tiny objects would be useless for such a purpose, and that they were simply the irreducible cores of cores and of no further use whatever; the larger, discoid cores are no more than they appear to be. Their technique was a simple one, designed to save the workman the trouble of preparing cores of tortoise type, and to enable continued flaking to be done until the core became too small. The method was to take a flattish piece, or a flake of raw material, and strike off sizable flakes all round the edge alternately. After a number of flakes had been removed in this way, the core might then be prepared for the removal of a single, large flake from one or both faces, and it was then that a striking platform was carefully worked on the edge of the core. This was not often done, however, and, more usually, the discoid core was flaked and re-flaked until it was discarded as too small for further use. The majority of the flakes detached in this manner have plain, Clactonesque, inclined platforms.

RUBBLE INDUSTRIES

Many Levalloisian industries are to be found all over Uganda, in various rubbly deposits. Most of the latter occur beneath recent red earths and the industries have every appearance of being rather late. They are almost always made in white quartz of poor quality, and only the presence of typical Levalloisian features, like the cores and flakes with prepared platforms, proves their cultural character. Such rubble industries cannot be dated with relation to stratigraphically fixed stages,

175

as the tools are too unspecialised, and they can only be judged in relation to such other tools as may occur in the same rubble deposits. Even this proves nothing more than that these Levalloisian stages are later than Upper Acheulean and earlier than Neolithic.

STILL BAY

In Europe, the Levalloisian appears to have died out or become merged with the Mousterian before the appearance of Neoanthropic man and his Aurignacian culture, but, in Africa, this was not the case, and the Levalloisian culture and its derivatives continued to exist well into Upper Palaeolithic, and even Mesolithic days.

From time to time actual cultural fusion took place between it and certain Neoanthropic, blade-and-burin industries, resulting in hybrid groups, in which both types are strongly represented.

In almost every Late Levalloisian derivative, one of the type tools is a point or lance-head, worked more or less all over by fine retouching, and this type of implement reaches its zenith in the Still Bay. Unfortunately, the frequent occurrence of certain Neoanthropic types in several Still Bay industries at various places has led to the assumption that the Still Bay *itself* was the result of contact between Late Levalloisian and Neoanthropic cultures; that is, by some unexplained freak of development, the plain, backed-blade-and-burin technique is supposed to have given the Late Levalloisian people the idea of making beautiful points and lances.

In my opinion, nothing could be more unlikely or further from the facts, as a perusal of the evidence soon shows. Points, sometimes delicately worked on one or even part of both faces, were common in Late Levalloisian industries before Neoanthropic influence appeared, and their development into still better shapes and with even more skilful retouching was a normal evolution which had nothing whatever to do with the tardy borrowing of such types as backed-blades, end-scrapers and burins whose technique was utterly different.

As I hope to show, there is a strong likelihood that the origin of points and lances in Levalloisian derivatives is to be sought in the contact between the Levalloisian culture and others in which pointed, *biface* tools were common, such as the Fauresmith of South Africa and the

Upper Tumbian of Uganda. These two groups are both of African Acheulean descent, though the Tumbian is both later and further removed from the parent stock and, if the idea of Levalloisian-plus-*biface* culture proves valid for the origin of the Still Bay, then we must recognise a multiple, immediate origin for it—with the Acheulean as a common ancestor—and, probably, for many kindred industries; and, further, that its birth certainly did not take place simultaneously in all parts of the continent.

As, however, the idea seems firmly established that the Still Bay is the result of a Late Levalloisian-Neoanthropic culture contact, it would be as well to review the evidence on this subject before going on to discuss the data from Uganda.

SOUTH AFRICA

The Still Bay was one of the first stone industries to be recognised in South Africa, but, for many years afterwards, it and a number of other related industries could not be properly placed in the sequence.

Before Burkitt's visit, in 1927, a whole group of industries, related to and including the Still Bay, were classed under the general term "Eastern Culture", but, at his suggestion, the nomenclature was revised, the name "Still Bay" reserved for the material from the type station, and it and all related industries included under the comprehensive term "Middle Stone Age". It was recognised that a strong similarity existed between the Middle Stone Age from South Africa and the Mousterian from North Africa, a similarity which is now known to be due to their common origin in the great, Old World Levalloisian culture of tortoise-cores and flakes with facetted striking platforms.

Unfortunately, the close connection between the Still Bay and other groups such as the Glen Grey Falls industry, which had originally led to their being classed together, seems to have been disregarded, or minimised when, after study of the Still Bay, Burkitt put forward his tentative opinion that this industry was in some way due to a contact between the Middle Stone Age and the "Neoanthropic blade and burin culture".[1] This was in spite of the fact that there had been obvious evolution towards the Still Bay in the earlier industries, before such

[1] *South Africa's Past in Stone and Paint*, 1928, pp. 88 and 170.

Neoanthropic influences became apparent. The following quotations make this clear:[1]

(1) *Glen Grey Industry*

Judging from the material collected from the Glen Grey Falls site...this industry shows primarily a flake technique.... The most symmetrical and carefully made implements...fall into a class midway between the small neat coups de poing *of the Fauresmith Industry and the beautifully made lances of the Still Bay.*

(2) *Pietersburg Variation*

The Glen Grey Falls Industry stands out very distinctly from the Still Bay, but an illuminating link seems to be present in the Pietersburg Variation. While on the one hand this may be regarded as merely a refined Glen Grey, yet on the other hand it shades into Still Bay forms, even equalling them in symmetry and beauty in a few instances.

(3) *Still Bay*

It has already been pointed out that the Pietersburg Variation seems to form a step from the Glen Grey Industry towards the Still Bay.... The typical implement of the Still Bay is again the lancehead, worked evenly and neatly over both faces.... The section is lenticular,...and as a result, the central keel so typical of the Pietersburg Variation is absent.

As regards the suggested Neoanthropic influence in the Still Bay, it is to be noted that, up to 1929, only a few specimens of crescents had been known and tentatively associated with actual Still Bay material. Most of these came from Fish Hoek or near by, and it was this association of typical Still Bay points with "Neoanthropic" tools that led Burkitt to postulate tentatively the contact which *resulted* in the Still Bay.[2]

(4) *Howieson's Poort Variation*

This industry appears to be definitely later than the Still Bay and, if a few tools, such as crescents, occur in the latter, there is nothing surprising in their occurring in large numbers in the Howieson's Poort group. In this stage, however, true burins make their first appearance,

[1] Goodwin and van Riet Lowe, The Stone Age Cultures of South Africa, *Ann. S. A. Mus.* vol. XXVII, 1929, pp. 104 *et seqq.*
[2] *South Africa's Past in Stone and Paint*, 1928, pp. 87, 170.

and the association is clearly due to something more than chance borrowing and seems to be a definite mixture of two distinct cultural facies. It is the first time that such mixture can be demonstrated as having occurred on an important scale, and the few crescents in the Still Bay cannot be regarded as more than elements borrowed from another culture that was just coming into contact with the Glen Grey-Pietersburg-Still Bay complex.

The sequence outlined above shows clearly that there was an almost perfect evolution of South African Middle Stone Age culture through the Glen Grey and Pietersburg stages towards the Still Bay. Throughout the whole group, the characteristic points can be seen gradually developing, becoming finer and receiving more and more retouch until, in the Still Bay, they are worked all over and even the facetted striking platforms are almost universally flaked away. The evolution appears to be quite normal and calls for no other explanation than the gradual improvement of a *biface* technique applied to flakes, so it seems unlikely that the presence of a few Neoanthropic elements in the Still Bay should actually have intensified the *biface* technique, already so well developed in the earlier stages. In fact, so far as the lances and points are concerned, it might well be permissible for the Glen Grey and Pietersburg groups to be regarded simply as early phases of the Still Bay and late stages of the Fauresmith. Such a derivation is not so unlikely as it may sound at first, for all these groups have much in common.

Speaking of the Fauresmith recently, Mrs Bowler-Kelley remarked[1] that it was *a combination of Micoque, Combe Capelle and Levallois characterized by Solutrean-like retouch and containing heavy blades and burins.*

The same possibility was recognised in South Africa at least as early as 1929, when Goodwin wrote:[2]

The 'dominant' (in the South African Middle Stone Age) would appear to be Mousterian (i.e. Levalloisian) influence first appearing with the Fauresmith and ousting the palaeo-anthropic coup-de-poing or at least reducing it in size to the lance head.

[1] A. Bowler-Kelley, *Lower and Middle Palaeolithic Facies in Europe and Africa*; paper read at Int. Symposium of Early Man, Philadelphia, March 17, 1937.
[2] Ann. S.A. Mus., vol. XXVII, 1929, p. 100.

And again:[1]

... These four groups (Glen Grey, Pietersburg, Still Bay and Howieson's Poort) *appear to show a sequence from the Fauresmith, leading through the Middle Stone Age to the Neo-anthropic industries.*

Sufficient has been said about the South African Middle Stone Age to show that the origin of the Still Bay points and similar tools was a good deal earlier than the time when a few Neoanthropic elements appeared in the industry. These were, at first, simply extraneous additions, in the form of lunates and, perhaps, backed-blades, but this foreign influence could not have modified tools already in existence to the extent of improving them up to the standard of the Still Bay points when the technique of the latter is that of the *biface* and utterly alien to that of the lunate or backed-blade.[2]

East Africa

Still Bay is abundantly represented in Kenya though mostly at open stations which are not home sites. In 1932, however, Leakey investigated the Apis Rock shelter, in Tanganyika, which yielded an interesting series of Levalloisoid industries. According to the brief account,[3] the succession began with

---

[1] Ann. S.A. Mus., vol. xxvii, 1929, p. 134.

[2] Since the above was written, *The Geology and Archaeology of the Vaal River Basin*, by Sohnge, Visser and C. van Riet Lowe has appeared. The following excerpts from Professor van Riet Lowe's account of the origins and nature of the South African Middle Stone Age are of interest:

It (Fauresmith II) *is a peculiarly interesting industry for the "type-fossil" is still the hand-axe, small, neat and beautifully trimmed in association with an advanced Levallois technique, well-trimmed points, blades... and the whole represents the beginning of the Middle Stone Age almost as much as it does the end of the Earlier.... It seems not to have... been entirely replaced by the Middle Stone Age but rather to have persisted into it... and we suspect moves and counter-moves and one culture borrowing from and blending with the other* (p. 113)*....the "type-fossils"* (of the Middle Stone Age) *are the Levallois tortoise core and a "point" on a flake often with facetted butt.... Throughout the sequence of industries, the core remains much the same, but the tools—flakes and blades—show a progressive development and refinement through a percussion to a pressure technique* (p. 114)*....a link between an advanced hand-axe culture with its percussion technique only and the Still Bay with its percussion and pressure techniques is something which I constantly expect* (p. 118).

[3] *Stone Age Africa*, 1936, p. 62.

...a pure and simple developed Levalloisian. Next comes a level which yielded very many implements of the type which I would formerly have called 'Upper Mousterian'. Associated with these are a few 'backed blades' and end scrapers, types of tool borrowed from the Aurignacian culture as a result of the contact. In this level there are also a number of specimens on which...the secondary trimming is extended over both faces of the 'points', and immediately above this occupational level is another, which yielded a typical Still Bay stage of culture.

In consequence, Leakey has revised the name Upper Kenya Mousterian, and changed it to Proto-Still Bay.

Judging by the excellent illustrations of this material, there is not much difference between the proto- and developed Still Bay groups, and the former seems to me to be distinctly better than the original Upper Kenya Mousterian stage, figured in the first Kenya volume.[1] Apart from the tools worked *all over* on both faces in the Apis Rock Proto-Still Bay, almost all the other points show some work on both faces, whereas, in the supposedly equivalent Upper Kenya Mousterian industry, in Kenya, only a few specimens show a little flaking on the lower surface, and, among the examples actually illustrated and so, presumably, the most typical, *none of them have any signs of work on the lower or main flake surfaces.*[2]

All these facts suggest that the Apis Rock Proto-Still Bay is an intermediate stage between the "Upper Kenya Mousterian", or Proto-Still Bay, and the upper, or developed Still Bay. In other words, that the beginning of the Early Still Bay is represented in Kenya, but not at Apis Rock.

In any event, the presence of a few backed-blades and end-scrapers in an industry already showing marked Still Bay characters cannot possibly be regarded as proof that the contact, if contact there was, led to the Still Bay. This is exactly the situation created in South Africa, when it was not realised apparently that Still Bay tendencies had already shown themselves long before there is any suggestion of "Neoanthropic" influence. I am convinced that, had that fact been appreciated, an unfortunate misconception would not have found its way into African prehistory.

[1] *Stone Age Cultures of Kenya Colony*, 1931.   [2] *Ibid.*, p. 86.

There is, of course, no doubt that contacts between the Still Bay and blade and burin industries have taken place, but, again, it is unlikely that they all occurred at the same time in every place. On the contrary, it is probable that, in some areas, which were early colonised by the blade and burin culture, the contact with Levalloisian and Early Still Bay industries was rapid, while in others, only the later stages of the blade and burin culture finally impinged upon the local Still Bay.

As examples of early contacts from which the Still Bay borrowed certain "Neoanthropic" elements, one may cite the Bambatan Industry of Southern Rhodesia and, possibly, the Early Still Bay of Kenya. Of instances of later contacts there are many, including the Still Bay of South Africa (which is here regarded as a late stage of the Glen Grey-Pietersburg, *Early Still Bay*, complex), Apis Rock and, much later, the Late Still Bay-Early Wilton contact that resulted in the industry known as Magosian in Uganda and in the upper levels at Apis Rock.

It has already been shown that there is more than a possibility that the Still Bay of South Africa is descended from the Late Acheulean-Levalloisian contact culture—the Fauresmith. Can this also be said for Kenya? I believe it can, although no certain evidence is forthcoming yet. There is, however, the fact that the Late Acheulean of Oldoway occurs right at the top of the sedimentary series whose deposition was truncated by earth movements and consequent erosion. Between the Late Acheulean ashy deposits and the rest, which are Gamblian gravels, containing Kenya Aurignacian tools, there is, obviously, some considerable gap, so we do not know what happened to the Late Acheulean during this period. However, on the Kinangop Plateau, overlooking the Kenya Rift Valley, quantities of these tools occur in deposits overlying ancient, pre-human, faulted Kamasian sediments. Personal examination of many exposures along the Kinangop scarp, in company with Solomon, showed that the earliest tool horizon is a rubble layer which occurs mainly in old erosion basins and shallow ravines cut back, in the Kamasian sediments, from the Rift edge. In some places, the rubble is hardened and reddened and it is quite clearly a typical land-surface deposit. A few tools of the type familiar around Nanyuki (that is, "Nanyukian" choppers and flakes) were seen, but left, in the rubble and there can be little doubt that this is the main tool horizon for the earlier material.

Filling in and levelling off the previous, rubble-filled depressions and shallow ravines, is a brown, clayey deposit which is clearly a sort of wash material and not, as claimed by Leakey, a Kamasian bed *in situ*, though it may be composed of washed Kamasian material. This contains an abundant series of beautiful little Still Bay points. I knew that these had been called "Pseudo-Still Bay" and were considered to be of Late Kamasian age,[1] so, while at the Nairobi Museum, I made a careful comparative examination of this and other Still Bay material. The result was amazement that the extremely slight differences between the Kinangop series and the developed Still Bay of other Kenya sites should have been interpreted as demanding so large a time-difference and the relegation of the former to a "Pseudo-Still Bay" industry.

Though we examined many scattered washouts, we never saw a trace of a stone, far less a stone tool, in true Kamasian beds, while the erosional unconformity between these and the implementiferous rubble may be of any age, so that it is quite impossible to date the Rift faulting by these deposits, as has been done.[2]

On the other hand, there is a possibility that the small, Late Acheulean hand-axes from the Kinangop Plateau are later than the Nanyukian tools, for those that I examined in the Nairobi Museum were usually fresher than the latter. We may, therefore, bear in mind the possibility that this Late Acheulean lasted into the Gamblian, where it may well have influenced the Kenya Levalloisian towards the Still Bay by transmitting to the flake industry its *biface* method of flaking pointed tools.

In making this suggestion, I am aware that Leakey is also inclined to a similar opinion, but he applies the reasoning to the Kinangop Still Bay,[3] and mistakenly, because the latter is clearly a fairly advanced stage, whereas the contact must have taken place at least at the end of the old Lower Kenya Mousterian, that is, at the beginning of the Lower or Proto-Still Bay.

In Uganda, the Still Bay is not at all common, in spite of the wide extent of the Late Levalloisian. I know of only one site where Still Bay

[1] *Stone Age Africa*, p. 42.　　　　　　[2] See footnote, p. 204.
[3] *Stone Age Africa*, p. 49.

of the Kenya facies occurs—that found many years ago by Mr W. C. Simmons of the Geological Survey, near Hoima, Western Province.

In eastern Uganda, however, near Jinja and round the foothills of Mount Elgon, there is a most interesting and prolific industry of Early Still Bay type, in which the *biface* tradition seems clearly to be derived from the Upper Tumbian. Unfortunately, at both these sites, the tools occur in a rubble in which there was but little trace of stratigraphical succession, but study of the relative states of preservation has enabled us to work out an age-and-typology succession reasonably close to what must have been the original sequence.

The first of these sites is the Bugungu Plateau, behind the Ferry, opposite Jinja. This laterite plateau has already been referred to in the section on the evolved Kafuan, where it was seen that the deposit was originally formed as a shore silt at about the time of the last south-west to north-east land tilt, just after the time of the Kagera valley Middle Tumbian and Levalloisian.

We do not know how long the high water-level, induced by this tilt, was maintained, but it was probably not for very long. When the level fell, after the lake found an outlet near the present Ripon Falls and caused deep erosion as the overflow surged away north to Lake Kioga, these shore deposits became a land surface in which three different series of roughly contemporary artifacts were subsequently entombed when the whole deposit became lateritised. The first of these occurs down to 2 ft., and is the Late Kafuan; the second, which is a developed Levalloisian, is to be found in many places, firmly adhering to the top of the laterite, while the third series, Upper Tumbian, very likely belongs to this period, for we found tools of this culture, with lateritic adhesions, lying on the surface or in rubbles.

Erosion of the laterite has resulted, in places, in the formation of a deposit of detrital fragments and tools from it. This detrital rubble also contains, with the older tools, a large group which must postdate the hardening of the bed as these implements are much less deeply stained and quite free from adhesions of laterite. The whole of this series, with the exception of the Late Kafuan, which did not continue, is divided into two groups—Upper Tumbian and Upper Levalloisian. Both these groups are redivisible into a number of successive stages according to their preservation and staining. By this method, they fall into four main classes, according to whether they were (1) very dark brown or purplish, with lateritic adhesions, (2) light brown, with some adhesions, (3) yellow, or (4) white or unstained.

These four classes were then subdivided into stages *a, b, c* and *d.* It was then seen that only Class I and part of Class II came actually out of the laterite, as these alone bore adhesions, while all the rest were later. That is to say, by the time of Class II, stage *c,* the Bugungu deposits had hardened so that no more tools became embedded in them. With the exception of the Late Kafuan material, there was no typological break between the laterite and post-laterite series, the tools only became progressively smaller and more finished in technique. This rule applies more strictly to the *biface* tools than to the Levalloisian cores and flakes, and it is interesting to see that, as time went on, the former steadily lost their ordinary Tumbian character and took on a distinctly Still Bay appearance. Finally, in Class III, stage *b,* the tools are exactly like the tiny *bifaces* of the Early Still Bay of the second site, near Mount Elgon, but, at Bugungu, the Tumbian is not entirely submerged, for true Tumbian *bifaces* occur as late as III*d,* that is, as late as developed Still Bay. The latter, however, is rare as yet, and the continued evolution of the group is better seen at the other main site, some 90 miles away.

WALASI VARIATION

Wayland had collected implements from the Walasi Hill neighbourhood before we went to Uganda and when we saw them in his collection we were at once struck by the association of tiny *bifaces,* like miniature hand-axes, with ordinary, if late, Levalloisian cores and flakes. At the time, we supposed that the industry was the result of contact between Levalloisian and Late Acheulean, but excavation soon showed that the little *bifaces* were really only early forms of the Still Bay point.

Owing to the lack of any but the slightest stratigraphy in the Walasi subsurface rubbles, the evolution of these points was not absolutely clear at first, although we noted that, in the majority of cases, the real developed Still Bay points tended to occur at or near the top of the rubble. One site, however, proved exceptionally valuable in establishing a clear succession based, once more, upon changes in the state of preservation and typology of the series, and it was evident that the site had seen some lengthy occupation.

This, known as the School Site, was on a flat hilltop, capped by rotted laterite and red earth, in which the tools occurred. In common with all objects in contact with such iron-charged material, the implements were variously stained, according to the length of time they had lain in the deposit. At all the other sites, the rubble was composed of rotted granite sand and quartz chips which produced little or no stain on the tools in it.

The following sequence was obtained from the School Site, all the material being quartz except for a little very fine-grained, pale quartzite:

4.   White (unstained); Neolithic conical cores, lunates, etc.

3*d*. Faint yellow; trace of Late Still Bay or Magosian.

3*c*. Light yellow; developed Still Bay, well-made points etc.

3*b*. Dark yellow   ⎫ Early Still Bay, crude points, small *bifaces*
3*a*. Light yellow-red ⎭   etc.

2*b*. Dark red   ⎫
2*a*. Red-brown ⎭ Indeterminate cores and flakes; Late Kafuan.

1.   1 Upper Acheulean hand-axe.

Technique and Typology

Taking the oldest first, this single implement calls for no comment; it is a small, well-made Upper Acheulean hand-axe.

Stage 2 (*a* and *b*) is an unspecialised industry in which there are no Levalloisian tools. It resembles, if anything, the Late Kafuan of the Bugungu laterite.

From the beginning of Stage 3, Levalloisian cores and flakes with prepared striking-platforms are common until Stage 4, the Neolithic. Almost all the cores are of discoid type and, in this, the group agrees with all other Uganda Upper Levalloisian industries, but at Walasi such cores were frequently struck twice, that is, from opposite ends, upon the same face. Numerous other core-tools, such as choppers, core-scrapers and, possibly, hammer-stones and sling-stones also occur.

Very few of the flakes were retouched and such as were usually took the form of rather poor side-scrapers.

Crude Early Still Bay points also make their first appearance in Stage 3. At first these are simple little *bifaces*, usually oval or slightly pointed. They are thick in comparison with their width and resemble nothing so much as extremely small, Lower Palaeolithic hand-axes. They can be seen becoming gradually thinner and narrower until, in Stage 3c, real Still Bay points and lances are the rule. The evolution is the same as at Bugungu and there is no doubt that the industry is the same also.

In Stage 3d, the industry exhibits a marked falling-off in quality, there being practically no points, but Levalloisian cores and flakes are still numerous.

While there are no true Tumbian artifacts at Walasi, as there are at Bugungu, I feel that the Levalloisio-Tumbian derivation of this industry is demonstrated by the typology of the points which are exactly like those from Bugungu, where the evidence in favour of this contact is very strong.

Owing to the poor quality of the raw material usually employed both at Walasi and Bugungu, few, if any, of even the best Still Bay points reach the high standard of the Kenya industry, made in obsidian, so the two groups are not strictly comparable, but, even apart from this fact, the typological data does not support the view that they have had the same origin. While, in the Uganda material, the Tumbian influence appears to be very marked, the Kenya Still Bay seems to have a distinct "Late Lower Palaeolithic" feeling and, in this respect, bears a close resemblance to the other Still Bay industries of the Rhodesias and South Africa. In my opinion, the Uganda Still Bay was an independent local development from the Upper Levalloisian by the picking up of the *biface* technique from the Tumbian. For these reasons, while retaining the term Still Bay, I propose to call the industries from Walasi Hill and Bugungu the Walasi Variation, because of its apparently distinct origin as compared to the Kenya and other Still Bay industries.

HOIMA STILL BAY

Typologically, this industry, found by Mr Simmons of the Geological Survey in the Hoima area, is far more akin to the normal East African

FIGURE 28

Levalloisian Culture

Nos. 1–5: Lower Levalloisian flakes with facetted striking platforms. From the Horizon A in the post-M-Horizon current-bedded sands (see p. 173). Nos. 1 and 2 are in white quartz, the rest are in quartzite, fresh.

Nos. 6–9: Middle Levalloisian cores, flakes and scrapers from Horizons D, E and F in the N-Horizon Bed (p. 173); quartzite, fresh.

$\frac{2}{3}$ scale.

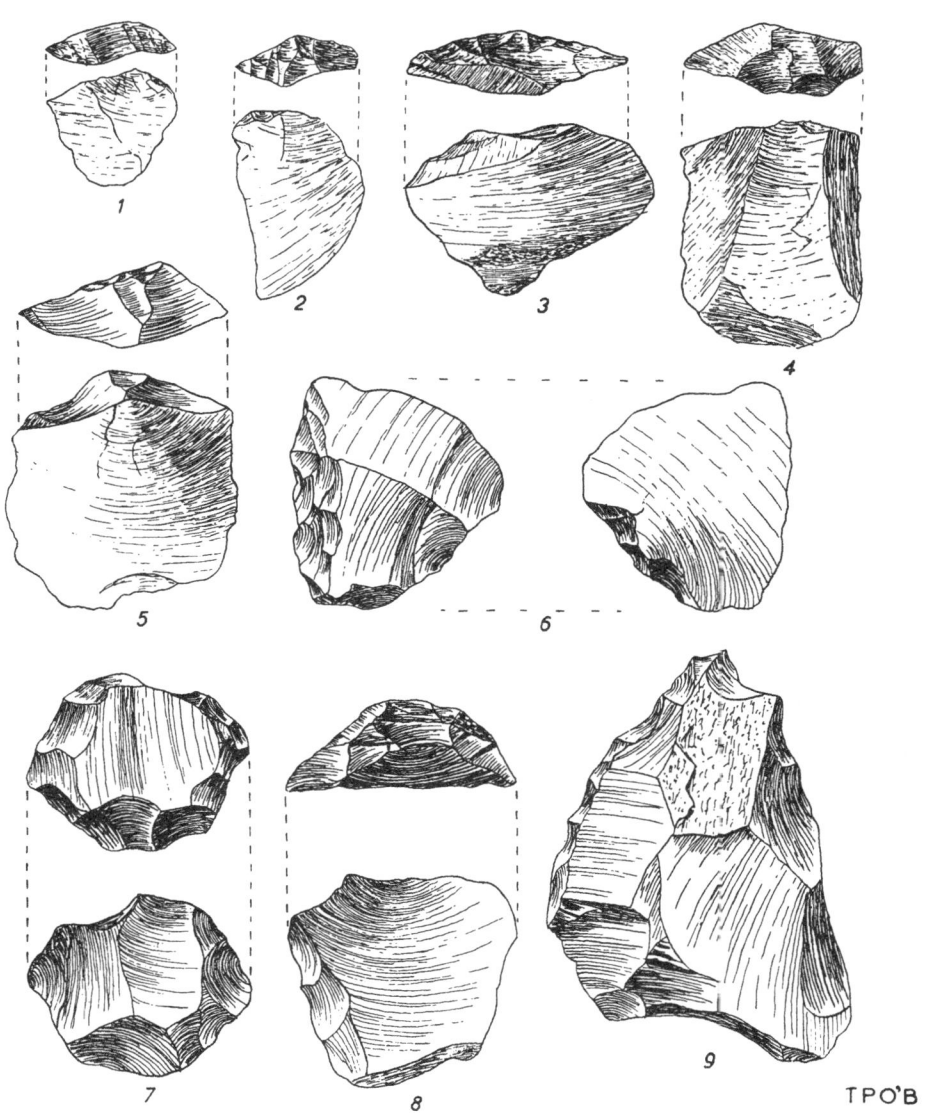

1   2   3   4   5   6   7   8   9

TPO'B

FIGURE 29

Levalloisian Culture

Middle Levalloisian tortoise cores from the N-Horizon Bed (see p. 173) ; quartzite, fresh.

$\frac{2}{3}$ scale.

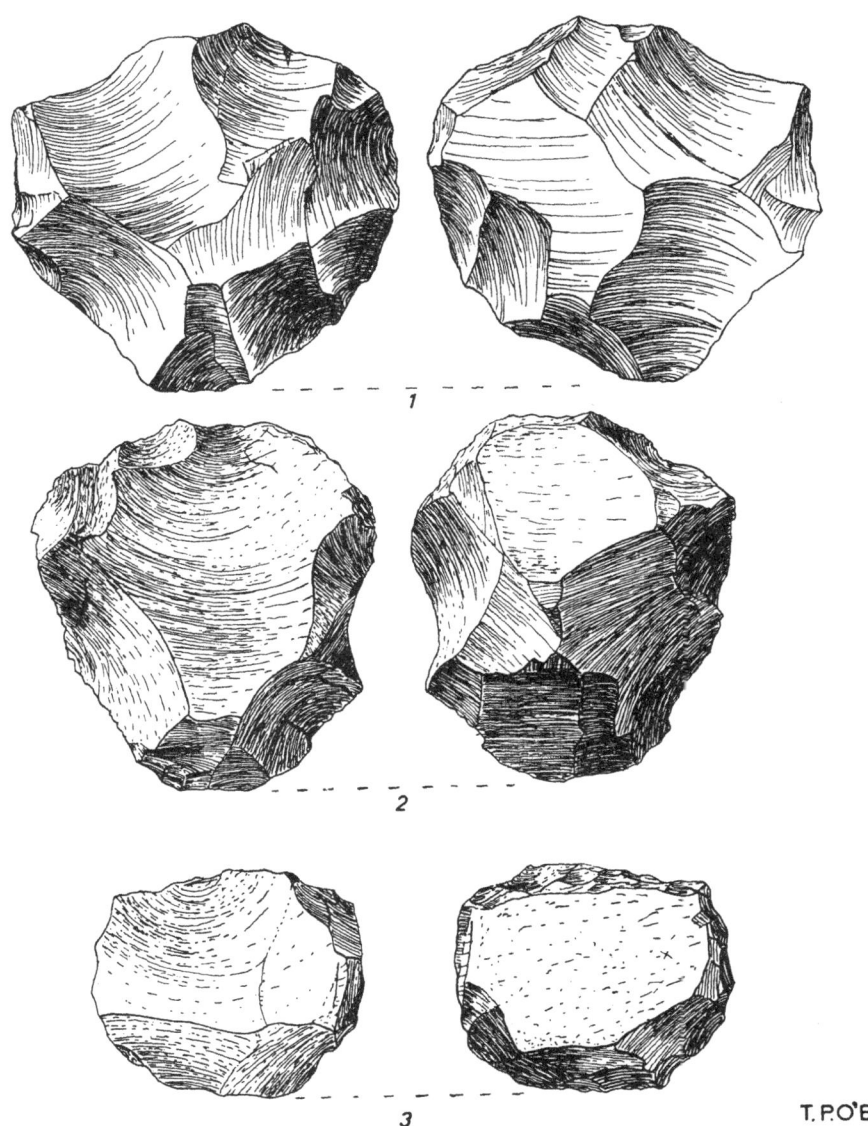

1

2

3

T. P. O'B.

FIGURE 30

Levalloisian Culture

Middle Levalloisian facetted-butt flakes and scrapers from the N-Horizon Bed. Nos. 1, 2, 3 and 4 are in quartzite; 5, 6, 7 and 8 are in white quartz, fresh.

$\frac{2}{3}$ scale.

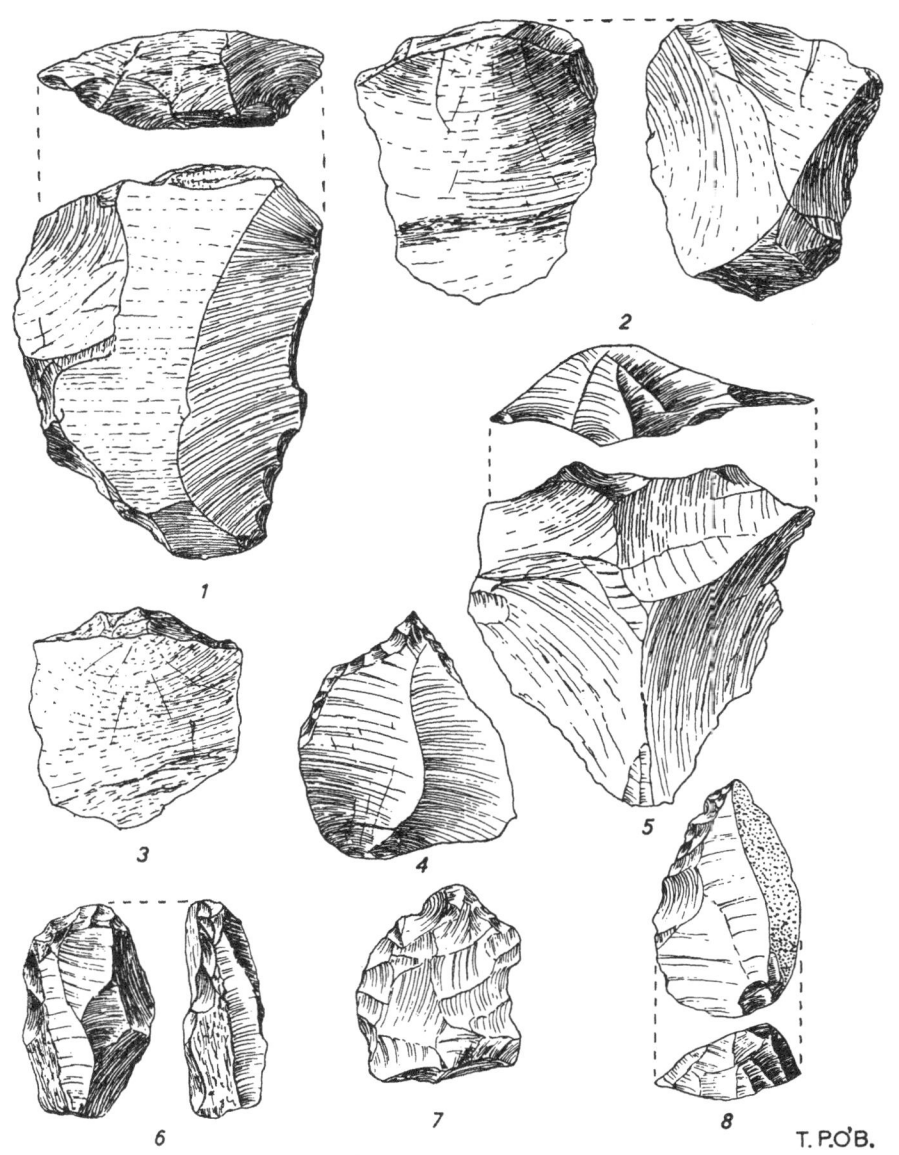

1

2

3

4

5

6

7

8

T. P. O'B.

FIGURE 31

Levalloisian Culture

Upper Levalloisian tools from a horizon in the post-100 ft. $\pm$ terrace, Orichinga valley alluvium, near Nsongezi. There is a slight feeling of Still Bay in this stage, as seen in No. 7. Nos. 1, 2 and 3 are in quartzite; 4, 5, 6 and 7 are in white quartz.

$\frac{2}{3}$ scale.

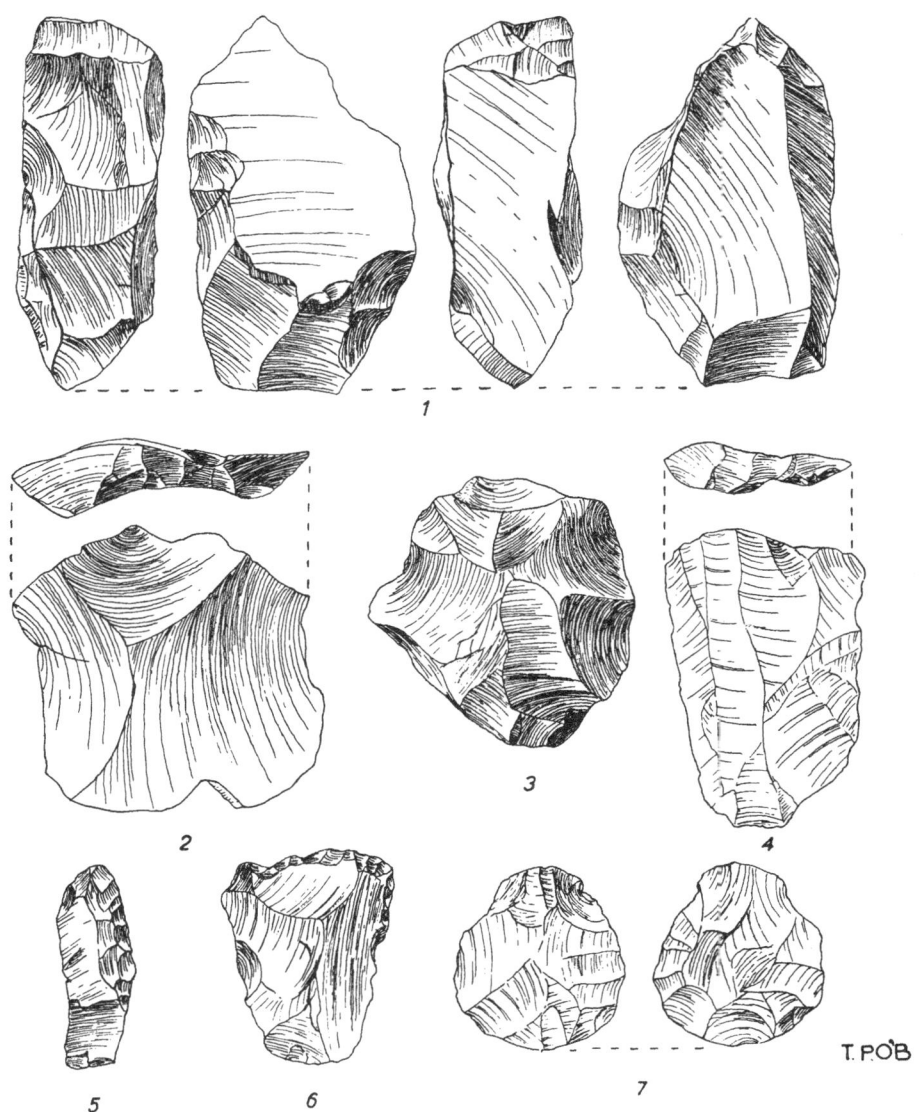

1
2
3
4
5
6
7

T. P. O'B

13·2

FIGURE 32

Levalloisian Culture

Tortoise cores of the Walasi Variation of the Still Bay from a rubble at Walasi Hill, near Mbale; white quartz, fresh.

$\frac{2}{3}$ scale.

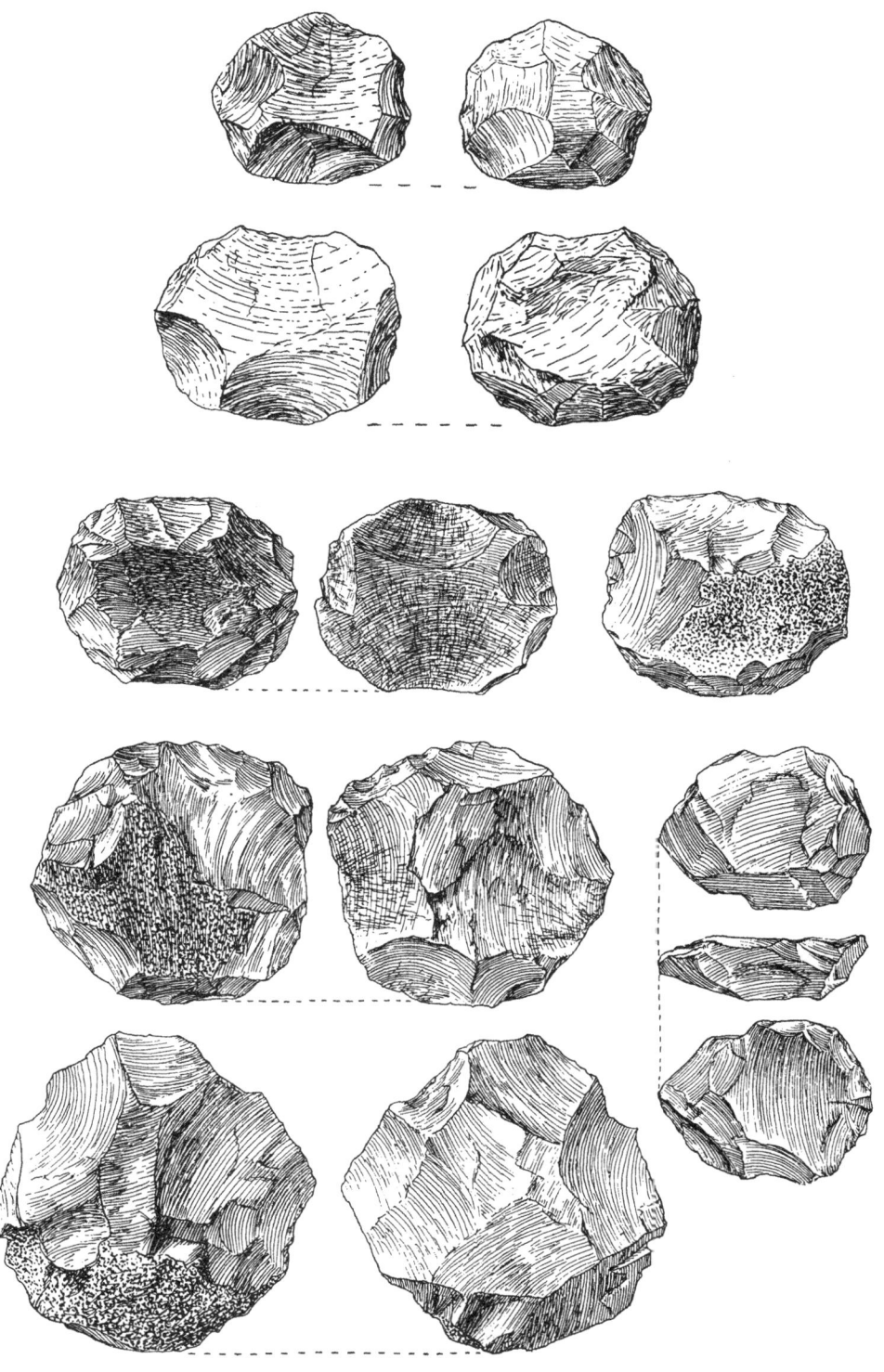

197

FIGURE 33

Levalloisian Culture

A series of points of the Walasi Variation of the Still Bay showing stages in the development of the leaf-shaped point from much larger *bifaces* whose appearance suggests descent from, or connection with, the Tumbian. From a rubble at Walasi Hill, near Mbale; white quartz, fresh or slightly weathered.

$\frac{2}{3}$ scale.

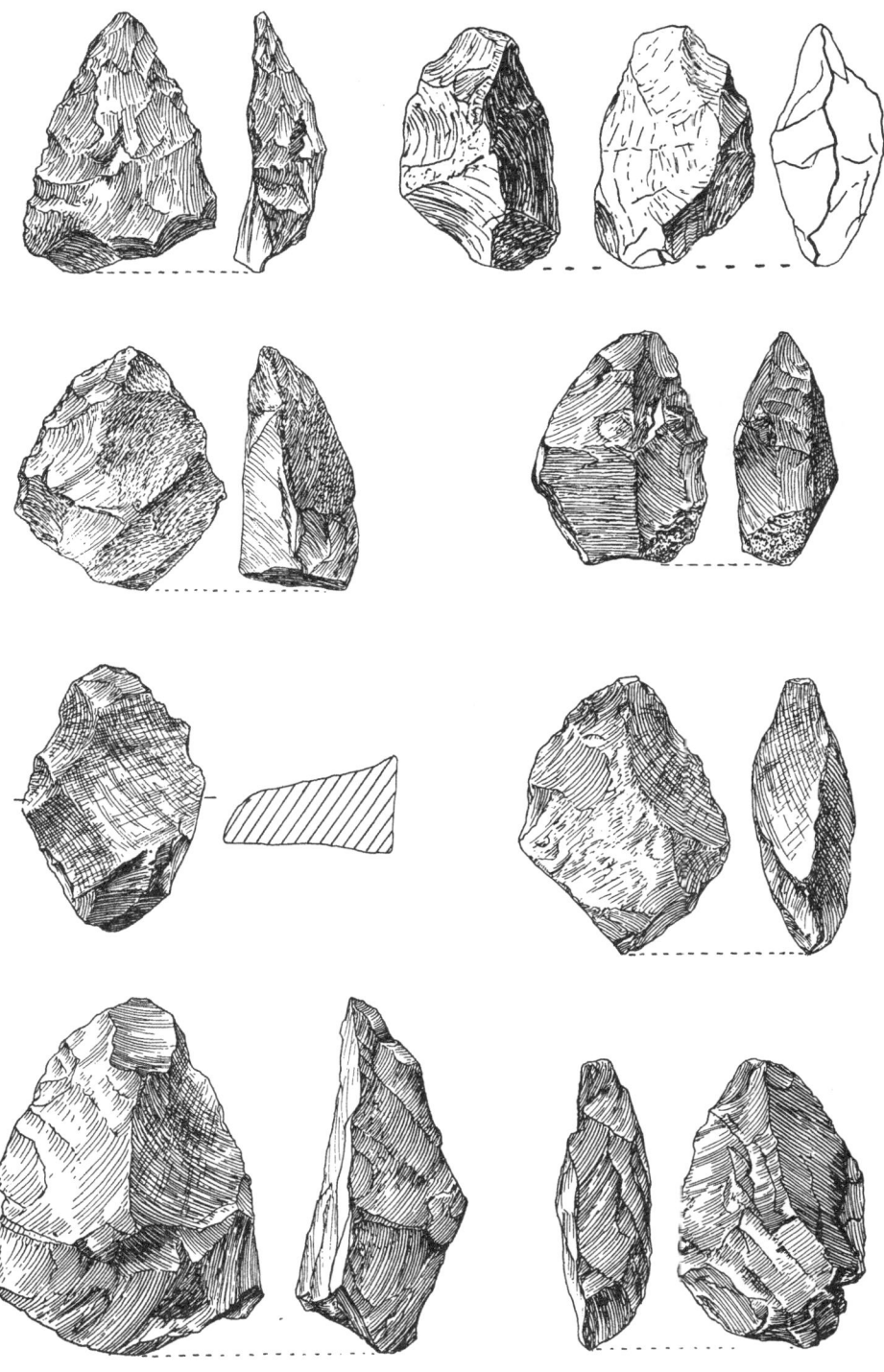

FIGURE 34

Levalloisian Culture

A series of points of the Walasi Variation of the Still Bay showing later stages in the development of the leaf-shaped point from larger *bifaces* whose appearance suggests descent from, or connection with, the Tumbian. From a rubble at Walasi Hill, near Mbale; white quartz, fresh or slightly weathered.

$\frac{2}{3}$ scale.

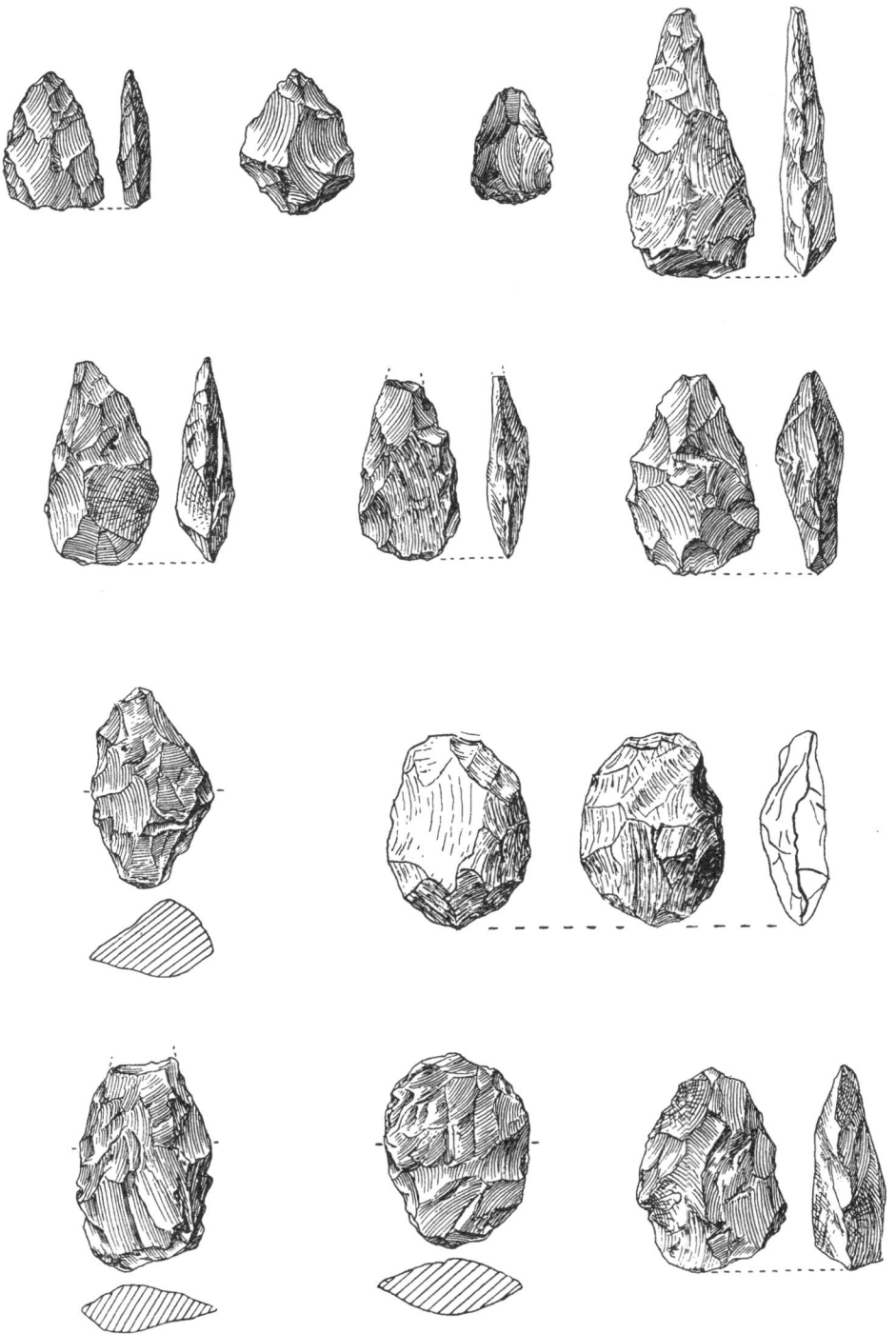

201

FIGURE 35

Levalloisian Culture

Facetted-butt flakes, flake-points, a chisel-ended tool and scrapers of the Walasi Variation of the Still Bay. From a rubble at Walasi Hill, near Mbale; white quartz, fresh.

$\frac{2}{3}$ scale.

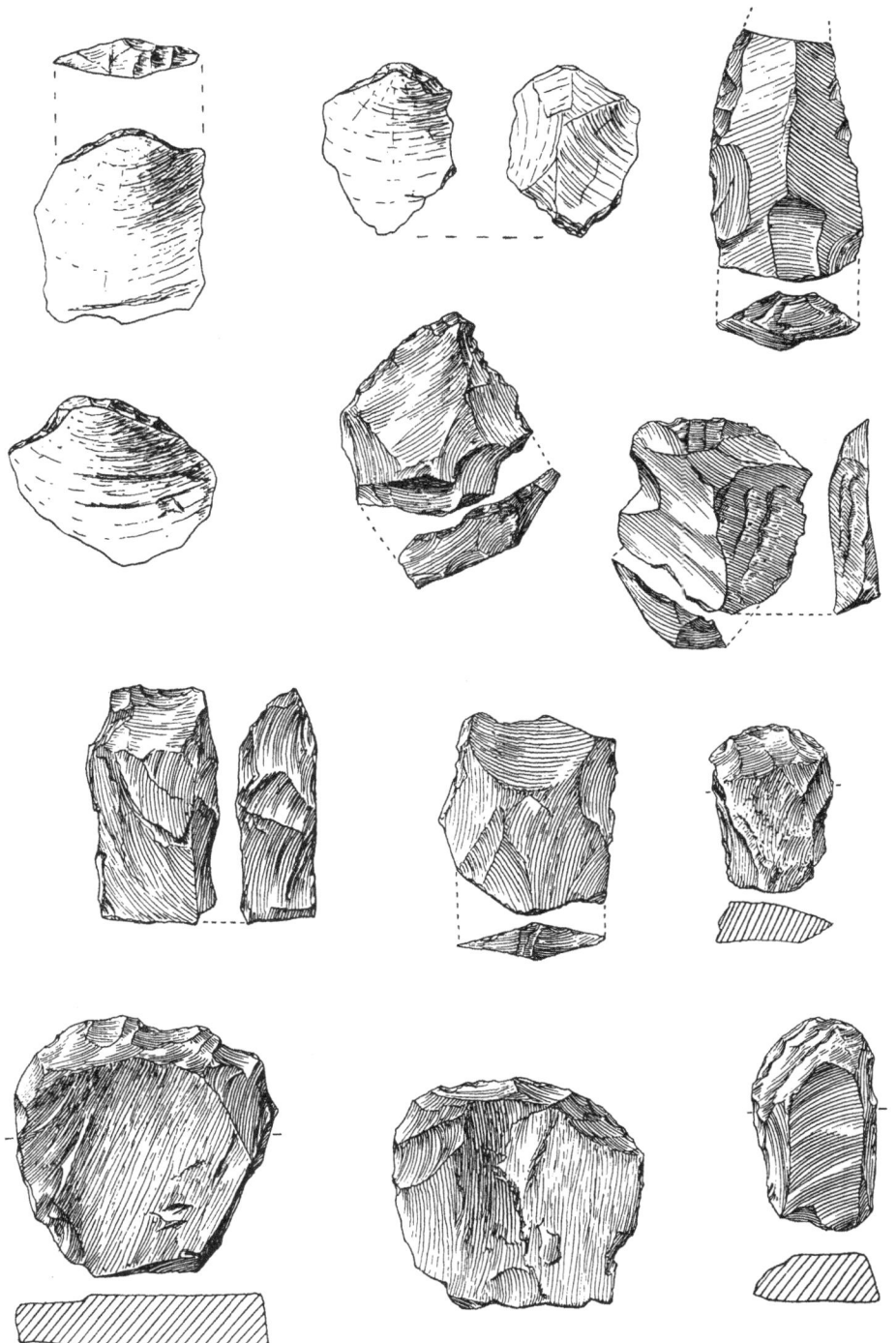

Still Bay than to the Walasi Variation. It is neither abundant nor very advanced and only occurs, as far as our own investigations showed, in a thin, poorly developed rubble.

The material employed—a black or grey variety of veined chert—made the manufacture of fairly symmetrical points much easier than when using quartz, as is usual in Uganda. These tools are, for the most part, oval or leaf-shaped and thin in section, with fine, parallel flaking over both faces; they are few in number, however, and the vast majority of the artifacts of this industry are simple Levalloisian cores and flakes, many of which have plain striking platforms.

Since I wrote this chapter, Dr Leakey has kindly conducted me to certain sites on the Kinangop Plateau rift edge, where he showed me tools *in situ* in what appeared to be waterlaid, stratified deposits.

If it can be proved beyond doubt that these are true swamp deposits and, further, that the tools have not found their way into them at a later date, the post-"Pseudo-Still Bay" age of this part of the Kenya Rift faulting is certain and my remarks on the subject, on pp. 170 and 182–3, are invalidated.

CHAPTER XI

The Uganda Tumbian

THE Tumbian culture is one of the most interesting in the whole of Africa for a variety of reasons, since it seems likely that it was a long-lived complex, descended from the Lower Palaeolithic and lasting until Neolithic times and, further, that it exercised an important influence on other cultures.

In spite of its importance in West and Central Africa, it has not yet received very much detailed investigation, while its enormous extent has tended to make the results so far obtained somewhat local and limited. As a result of this, a host of Tumbian industries have been recognised all over tropical West Africa which, at first sight, appear almost hopelessly incapable of synthesis. Actually, however, despite its widespread character, the Tumbian culture seems to have preserved its special characteristics almost everywhere and it should be possible to trace its evolution with less difficulty than that of many other, less specialised cultures.

Menghin was the first to realise the importance of this culture, to give it its name and to study it systematically.[1] From the typology of some of the tools, he came to the important conclusion that it was descended from the Lower Palaeolithic and was, at least partly, of Upper Palaeolithic age in the Congo. Owing to an unfortunate correlation which he effected between it and the European Campignian, because of certain supposed resemblances, the idea of its descent from the Lower Palaeolithic received scant support from other workers for some time. Added to this was the fact that, at many Congolese stations, polished tools also occurred, giving it a Neolithic appearance.

Judging by the published material and the few museum collections that I have seen, the Neolithic age of a good deal of the Central African

---

[1] O. Menghin, Die Tumbakultur am unteren Kongo und der Westafricanische Kulturkreis, *Anthropos*, tome xx, 1925, and in *Anthropos*, tome xxi, 1926. The culture was named after the station of Tumba.

Tumbian seems established, but, recently, the possibility of an earlier date for part of this culture has been admitted. After study of an industry from the Mouka plateau, Ubangi Shari, Breuil[1] came to the conclusion that the series showed an evolution from the Acheulean, and Mrs Alice Bowler-Kelley, after studying a large collection at the Musée du Congo Belge, at Tervueren, near Brussels, has expressed the same opinion.[2]

Menghin's conclusions on the morphology and typology of the several phases of the Tumbian appear sound and reliable. Breuil's remarks in this connection are appropriate here:[3]

Menghin est porté à penser qu'à la période plus ancienne du Tumbien se rapporteraient les gros outils pseudopaléolithiques et les feuilles de laurier, et que les pointes de flèches se multiplieraient dans la seconde partie. Comme exemple du facies le plus grossier, on peut citer le gisement de Kunzulu (Stanley Pool) découvert par M. Pring, où les outils volumineux et irréguliers sont à taille biface, passant de l'ovoïde globuleux aux types très allongés genre pic; mais la face inférieure est presque plate, tandis que la supérieure est très bombée.... Menghin pense...que cette civilisation serait dérivée du Paléolithique ancien et serait d'âge paléolithique supérieure.

EAST AFRICA

From material at the Museum of Archaeology and Ethnology, at Cambridge, given by the Ven. Archdeacon Owen, of Kavirondo Province, Kenya, it was clear, in 1933, that the Tumbian, or a culture extremely like it, had existed in Kenya. As it occurred exclusively in the west of Kenya Colony, it seemed very likely that this culture must also have been present in Uganda.

The preliminary examination of Wayland's material, in 1934, at once showed that it did occur. Unfortunately, all the Tumbian material that we saw was from surface sites, though it transpired, much later, that Wayland had obtained a few of these tools *in situ*. However, there was enough data to indicate that Menghin's suggestion of a Lower

---

[1] H. Breuil, *L'Anthropologie*, tome XLIII, 1933, p. 222.

[2] A. Bowler-Kelley, *Lower and Middle Palaeolithic Facies in Europe and Africa*; paper read at Int. Symposium of Early Man, Philadelphia, March 17, 1937.

[3] H. Breuil, Afrique, *Cahiers d'Art*, 1931, p. 94.

The Uganda Tumbian

Palaeolithic origin for the Tumbian was probably sound and, as the typology of the tools seemed to show, that an Upper Palaeolithic date for the earlier Tumbian was also probable.[1] At that time, however, we thought that the Acheulean derivation was a direct and simple evolution. It seemed that true Acheulean hand-axes grew progressively finer—thinner and narrower—until the tools became the typical, long *feuilles de laurier* and parallel-sided shapes. This view was quite wrong, however, and was responsible for some confusion in the field later on.

UGANDA PROTO-TUMBIAN

In the last chapter, I gave a detailed sequence of the post-M-Horizon deposits of the 100 ft. terrace in the Kagera valley, and it will be remembered that the immediate, post-M-Horizon false-bedded sands, containing Levalloisian, were followed by the N-Horizon rubble and its associated sands and clays. Apart from the M-Horizon, this is the best implementiferous layer in the 100 ft. terrace.

The N-Horizon and its succeeding sands with clay intercalations were laid unconformably on the Lower Levalloisian, false-bedded sands, and it is clear that, between them, there had been a period of subaerian erosion. In fact, in some places, the whole of the pre-N-Horizon sands down to the level of the M-Horizon had been removed and, in these cases, the N-Horizon rested directly on the M. These facts suggest that, after the deposition of the Lower Levalloisian sands by a sluggish Kagera and its main tributary, the Orichinga, a period of rather dry conditions supervened, the culmination of which was during the formation of the N-Horizon stone-bed, when the valley bottom was a dry land surface. The fine, white sands that cover the stone-bed probably represent the period of slow, climatic adjustment back to normal. Whether this dry oscillation had any effects outside Uganda is a matter for further work, when it may, perhaps, be found to equate with the Mid-Gamblian oscillation in Kenya.[2]

---

[1] T. P. O'B., Prehistory and Uganda, *Uganda Journal*, vol. II, part 3, p. 183.

[2] This oscillation caused the fall of the Lower Gamblian lake-levels and the formation of loamy sand and rubbly beds. See Solomon, *Stone Age Cultures of Kenya Colony*, 1931, Appendix A, p. 248.

Technique and Typology

As remarked in the previous chapter, the N-Horizon contains two quite distinct cultures, one being a substage of Levalloisian (the earliest part of the Middle Levalloisian) and the other the Proto-Tumbian. Further, whereas the latter is confined to the basal stone-bed, there were two more levels of Levalloisian in the succeeding bed of sandy clay. This fact, and the complete absence of any traces of contact between the two cultures in the N stone-bed, is proof that there was no connection between them in any cultural sense. Though certainly contemporaries for a time, it is clear that the two cultures had parted company in the Kagera valley by the time of the N sands and clays, so that the early Tumbian occupation of the area would appear to have been of comparatively short duration, a fact that may have been influenced by Levalloisian opposition.

Disregarding the rough flakes, a few of which were retouched as scrapers, the Proto-Tumbian tools are nearly all large, clumsy, stone-flaked hand-axes and picks. A number of small cores and core-choppers also occur. Almost all the hand-axes and picks are flat-bottomed, with thick humped backs. There are also a number of real, base-keeled tools, as in the Lower Acheulean, and the presence of these, and, indeed, the extremely crude nature of the whole industry, is probably partly due to the quality of the favourite raw material. This was a light grey, rather coarse-grained quartzite of tabular type, tending to break up into angular pieces. Very little use was made by the early Tumbian people of the far better, local, fine-grained, blue quartzite, although this was freely employed by the contemporaneous Levalloisian people.

The majority of the hand-axe tools tend to be elongated as well as thick, and might really be described as picks. The pick was one of the favourite tools in the Tumbian, both in Uganda and in the Central African branches, where it persisted until Neolithic days.

True hand-axes of Acheulean tradition are rare in the Proto-Tumbian though abundant later in the Middle Tumbian. This, at first, led me to think that the N-Horizon stage was a very degenerate stage of local Upper Acheulean, which somehow took on a new lease of life in the next stage, which I thought was Lower Tumbian, developing from Late

Acheulean[1] (the probable Gamblian age of these beds had not then been realised either, and we were working on the assumption that the whole complex was of Upper Kamasian age).

That idea, of course, was wrong. The N-Horizon material itself is Proto-Tumbian for, even though there are in it none of the tranchets and laurel leaves characteristic of the later Tumbian, the picks form a connecting link. Moreover, it is most unlikely that this rough assemblage could be a degenerate Upper Acheulean when there is so much well-made Upper Acheulean in East Africa, and the Proto-Tumbian industry is far more convincing when regarded as the beginning of a new culture than as the degenerate end of an old one. Apart from this, the presence of a well-developed Levalloisian makes it almost certainly post-Acheulean and of Gamblian age. At the same time, the considerable difference between the Proto-Tumbian hand-axes and those of the Middle stage is easily understandable when we regard the whole culture as a newly born one, in which, though the Acheulean tradition was present from the start, there was time and room for evolution from a crude beginning to the beauty of the later stages.

For these reasons, I find it difficult to believe that the Tumbian was the direct descendant of the pure Upper Acheulean, unless the latter had first reached a very degenerate stage, of which there is no evidence so far, in Uganda. A considerable period must have elapsed between the pure Upper Acheulean of the Lower Palaeolithic and the Proto-Tumbian of the Upper Palaeolithic, during which time most of the Acheulean culture either died out or was reborn as an industry such as the Fauresmith of South Africa. It is not difficult to imagine that, somewhere, a small, isolated branch or branches of the Acheulean escaped extinction, but only at the price of degeneracy, and it is from this sort of "surviving Acheulean" (later than the Upper stage) that I believe the Tumbian to be descended.

Apart from Uganda and the extreme west of Kenya, there is, so far, no sign of the Tumbian in East Africa and so those areas seem to have been on the periphery of the great Tumbian region, comprising Central and West Central Africa. Indeed, that peripheral zone serves effectively to divide Upper Palaeolithic Africa into two distinct, cultural regions—

[1] *Man*, 53, 1936.

the Central and West African Tumbian and the East African region of "blade and burin" culture.[1] Neither in Uganda nor in Central Africa is there any sign of the latter (except late derivatives).

I have already referred several times to the "Sangoan" and have described how previous ideas about its age and cultural character were based on misconceptions. Owing to the provenance of the tools, stratigraphical data is not forthcoming, so only a brief description will be given here, as similar implements were found by us in bedded deposits whose description has been fully given (N-Horizon).

In one of the early references to the "Sangoan", Wayland wrote:[2]

The tools...were all collected in the vicinity of Sango Bay, from the quartzite hills both north and south of the mouth of the Kagera River.... The soil is as a rule about three feet thick, though depths of six feet have been registered; in some places, however, there is no soil at all....Artificial excavations, such as trenches dug during the Great War, and a few small holes which I caused to be sunk for the purpose of tracing an outcrop, show that tools, flakes and cores are distributed throughout the soil from top to bottom. It must be remembered, however, that the ground has been much disturbed by cultivators.

These observations are entirely in accord with our own and, as I explained before, the only possible method of deciding whether more than one stage is present is to compare the various states of preservation. Our own collection showed that the bulk of the "Sangoan" material is only slightly weathered, while a few rare, more heavily weathered pieces belong to the Acheulean, mainly Upper. The large flakes clearly form part of the various stages and, except when they are obviously Levalloisian (in which case, they are usually small and fresh), are of little cultural significance.

The bulk of the more or less youthful "Sangoan" material is composed of two industries which are easily distinguishable. One is strikingly like our N-Horizon Proto-Tumbian; all the same tool-types are present—picks and coarse, hump-backed hand-axes with primitive stone flaking, core-choppers, cores and many plain flakes. The other is

[1] Levalloisian was, of course, ubiquitous.
[2] *Occasional Paper No.* 1, Geol. Surv. Dept. 1923, E. J. Wayland and Reginald A. Smith.

The Uganda Tumbian

Levalloisian and the whole assemblage is exactly comparable to our complex from the N-Horizon rubble.

A younger series is also present, consisting of Middle Tumbian and Levalloisian types like those of the uppermost clays that postdate the N-Horizon series in the Kagera valley.

Considering the immense quantities of "Sangoan" material on the hills in that area, it may yet be possible to obtain these several stages in stratigraphical succession there, or, at least, to work out a more detailed sequence, based on *état physique*, than has been possible so far. It may, for instance, be possible to subdivide the Proto-Tumbian and trace it back to its ultimate origins. In any event, it is clear that a great deal of research is waiting to be carried out on the Tumbian culture, both in Uganda and Kenya.

MIDDLE UGANDA TUMBIAN

In Uganda, the Tumbian reached the peak of its cultural achievement, in variety of form and beauty of workmanship, during the Middle Tumbian stage of the Kagera 100 ft. terrace.

As I said earlier in this chapter, the Proto-Tumbian occupation of the valley appears to have been a transitory affair, after which the makers of this industry moved away, either eastwards or back again to the Congo. Thus, the Middle Tumbian represents a fresh wave of invasion, after sufficient time had elapsed for the culture to have become much more advanced than the Proto-Tumbian stage of the N-Horizon stone-bed.

The deposit which follows the sandy clay above the stone-bed is, almost invariably, a single, thick, homogeneous grey clay. Towards Nsongezi (traced down the Orichinga valley to its junction with the Kagera), this deposit becomes more arenaceous and lighter in colour. Locally, there is an intercalated band of reddened sand or a zone of reddened clay which, on weathering, becomes hard. This whole deposit exists up to the top of the 100 ft. terrace.

When the last land-tilt took place, the 100 ft. terrace deposits were cut through by the river on its way to grade with the new Victorian base-level. Since that time, there have been two small cycles of gully-erosion in the valley floor, the second of which is in operation to-day. The earlier of these caused the excavation of long, broad ravines, entering the present valley obliquely.[1] It is probable that they were cut by storm waters which flowed down from the hillsides before the excavation of the present, main channel was completed, that is, when the drainage flowed across the then valley floor on its way to join the main Kagera. These ravines are now choked by dense bush.

ORICHINGA VALLEY CHANNEL

FIG. 36. Diagrammatic sketch of part of Orichinga channel
showing the two types of gully erosion.

A, shallow, usually long, bush-filled ravines.
B, steep-sided, short, open, irregular basins. Both types are cut in Kagera 100 ft. terrace deposits.

The second cycle of erosion is of recent growth and has resulted in the carving out of vertical-sided, narrow ravines entering the present channel at right-angles. It is the typical form of erosion to be expected in a region of uncertain climate and in deposits such as those of the upper part of the 100 ft. terrace. Very often, the upper ends of these ravines

[1] Though all the deposits considered here have been referred to as belonging to the Kagera, for the sake of simplicity, they were actually studied up the Orichinga valley, a mile or two from its junction with the Kagera.

have been widened into broad basins with flat floors, by lateral erosion due to the alternate action of storms and the drying and crumbling of the sides. The narrowest part of such erosion areas usually coincided with the lip of the terrace.

Eleven recent erosion basins of B type (see p. 212) were found, and quantities of Middle Tumbian tools were present in all of them. As these implements were nearly all mint-fresh and were so numerous, we were, at first, confident of finding the industry *in situ*. Pits and trenches were dug in the clays wherever we found a concentration of artifacts, but with very little success. Some tools were found in the clay, among which were a few typical Tumbian pick-hand-axes and some definite *coups de poing*, but none of the characteristic laurel-leaves or tranchets were found *in situ*.[1]

We thought that the reddened zone of sand or clay, visible in a number of sites, might be the implementiferous horizon,[2] but investigation proved this to be devoid of artifacts and we finally realised that the tools occurred sporadically throughout the clay, presumably having been dropped during visits to what must have been a passable marsh at that time. Only in a single instance did an intercalated, sandy layer in the Tumbian clay yield anything, but these objects were only a few poor cores and flakes, of no definite cultural significance. So the search for Tumbian horizons in the clay was abandoned, and we were forced to accept the rather unsatisfactory fact that the tools occurred all through the clay. This was unfortunate, as it prevented any morphological study of the Middle Tumbian, owing to the absence of stratigraphy. This is not a very serious matter, however, as the collected tools do not suggest that more than one stage was present.

Technique and Typology

The following types make up the Middle Uganda Tumbian assemblage:
    Picks, hand-axes, cleavers, large and small tranchets, oval, pointed,

---

[1] Wayland found some in place, when pitting the 100 ft. terrace, some years ago. I had no knowledge of this until recently. My thanks are due to Mr Wayland for permission to figure one of these specimens (see p. 253).

[2] The O-Horizon referred to in *Man*, 53, 1936. See also Solomon's remarks on pp. 32–33.

parallel-sided and tapering varieties of leaf-shapes, points, scrapers, hammer-stones, core-choppers, cores and flakes. The tools are made in blue, and grey quartzite and, occasionally, white quartz.

The picks are still mainly rather rough and clumsy and obviously intended for heavy work, being long and, usually, thick.

There are several characteristic hand-axe forms; one is a medium-sized tool with a steep, thick butt and a finely flaked, flat tip, pointed or rounded; the under side is usually flaked flat. The profile outline of these tools is rather like that of a duck's head. The second type also has a well-made tip, but its butt is rougher than that of the previous form, very asymmetrical and always had a flat area on it somewhere, from which flakes were struck to form a kind of steep scraper. The third class was of flat, well-trimmed tools with rounded butts and pointed tips, varying considerably in size. The smaller implements of this class closely resemble the cordiform hand-axes of Europe, and, indeed, none of the tools from this group would look out of place in any well-developed Upper Acheulean industry.

The rare cleavers are made in the usual "Lower Palaeolithic" fashion by retouching the sides of large flakes, leaving one edge un-chipped to form the blade, but they are not of the Victoria West type, for none of them have the parallelogram section and the bulbs are usually at the end instead of at one side.

The tranchets are typical Tumbian implements; they possess wide, untrimmed blades and narrower, thickish hafts, carefully trimmed on both sides and with an oval or round section.

The other, very characteristic Tumbian tools are the leaf-shapes, of which there is a great variety. Some are thin ovals and other elongated ovals with rounded or pointed ends and fairly thick in section. There are spear-shapes, long and wide and almost incredibly thin, considering that they are made of quartzite, and parallel-sided tools with rounded ends and points, while every gradation between all these forms is also present, all beautifully worked all over.

A number of points occur, some rather rough, others long, thin and finely flaked, also a few, rare scrapers, usually made on chunks with one edge retouched.

Hammer-stones and core-choppers were not very numerous, but cores and flakes were plentiful; some of the cores and core-choppers were made from boulders.

Almost every tool, except the picks, was made on a flake and the retouching, again with the exception of the picks, was by wood. The striking platforms that are visible—most of them were flaked away—are invariably plain and the whole group shows no sign of contact with the Levalloisian.

UPPER UGANDA TUMBIAN

Apart from the rubble collections from Bugungu plateau, opposite Jinja, we know of no Tumbian industries bridging the apparently long gap between the Middle Tumbian of the Kagera valley 100 ft. terrace and an obviously much later form, also found in that valley. The lack of stratigraphical evidence at Bugungu makes it difficult to fix precisely the date of the Tumbian there, but there is a strong probability that the evident evolution[1] of the Bugungu Tumbian culminated (locally) in contact with the Late Levalloisian and produced the Early Still Bay. That would make the Bugungu Tumbian of Upper Gamblian age and later than the Middle Tumbian of the Kagera valley, which is what its typology would lead us to expect. This dating receives geological support from the fact that the Bugungu material just postdates the 150 ft. $\pm$ beach, which was caused by the flooding of the north end of Lake Victoria following the last tilt and just before the Ripon outlet was established. This episode postdates the deposition of the Middle Tumbian clays in the Kagera valley.

Although the Bugungu data indicates an Upper Tumbian-Upper Levalloisian contact of considerable importance, it was probably no more than a local instance of the borrowing of a *biface* technique by the Upper Levalloisian, and I do not think that the local Tumbian was actually submerged. While there is but little further sign of it at Bugungu after the appearance of the Early Still Bay, this may be because the Tumbian tribe simply moved away from that immediate area; their culture certainly continued unaffected elsewhere.

[1] Judging by the changes in the various *état physique* classes.

215

Late Uganda Tumbian

From this point until the next appearance of the Tumbian culture, there is a gap in the sequence, for the final stage is, almost certainly, very late. It is best represented in the Orichinga valley, at one spot, near Nsongezi, and in such numbers as to make it certain that the place was an occupation-site.

Although none of the tools were found in place, their absence on the surface of the 100 ft. terrace suggests that they probably occur in, or just below, the brown soil that mantles the valley bottom. The tools in question were found in a wash-out, mixed up with Middle Tumbian artifacts that had eroded out of the 100 ft. terrace clay. They are small and always made of white quartz. Oddly enough, tranchets were by far the most numerous tools, the proportions being as follows:

Tranchets	71%
Points	18%
Crescents	1%
Hammer-stones	3%
Cores	7%

Other Late Tumbian tools were found sporadically in western Uganda, mostly between Kaiso and the Muzizi bridge, and a small tranchet was found at a granite rock-shelter, in a Magosian level.

Conclusions

1. Menghin's original claim, that the Tumbian was of Lower Palaeolithic descent, is considered as proved.

2. Similarly, the Upper Palaeolithic age of the Tumbian is established with reasonable certainty.

3. The derivation of the Early Tumbian was not directly out of a pure Upper Acheulean stage but, more probably, from a very late, degenerate stage which had escaped extinction and was itself of Upper Palaeolithic age.

4. The homeland of the Early Tumbian was probably West or Central Africa, whence the culture spread eastwards. Two waves of the culture reached East Africa, the first of which seems to have been of

short duration, while the second—of Middle Tumbian age—stayed on and continued its development.

5. In Upper Tumbian times there was probably contact between this culture and Upper Levalloisian which led to the birth of a local form of Still Bay, which is given the name Walasi Variation.

6. The culture continued to develop until, in Uganda, it died out in Mesolithic times, that is, before reaching a polished stage. Small local contacts with other cultures and industries probably took place.

7. The Tumbian is essentially a West and Central African culture. It does not seem to have spread farther east than Uganda and the extreme west of Kenya. That peripheral zone serves to divide Upper Palaeolithic Africa, south of the Sahara and north of the Kalahari, into two distinct cultural regions: the Central and West African Tumbian and the East African blade-and-burin areas, though there may have been slight easterly penetrations, as into Southern Rhodesia and else-where.[1]

8. Finally, it is not improbable that the Tumbian was responsible, in some way or another, for the diverse *biface* industries of the Sahara and North Africa, such as the Sbaïkian and Aterian.

[1] For instance, Prof. van Riet Lowe has drawn attention to certain similarities between some Fauresmith types and others, obviously Tumbian, from the Congo. This *may* point to infiltration of the Tumbian culture into South Africa. "Notes on the Resemblance of certain Stone Implements from the Belgian Congo...to Fauresmith types." *S.A. Journ. Sci.* vol. XXXII, November 1935.

PLATE XV

Tumbian Culture

Three views of a Proto-Tumbian artifact, probably a pick, with a pronounced base-keel on the underside which is unworked except for the removal of a single flake from the tip. As in almost all cases of base-keeled tools, the two faces on either side of the keel are flat, weathered joint or cleavage planes in the rock. From the Kagera valley Younger Rubble, near Nsongezi; quartzite, very slightly weathered. Repeated on p. 225.

Approx. $\frac{1}{2}$ scale.

PLATE XVI

Tumbian Culture

Fig. 1. Two views of a Proto-Tumbian pick from a rubble on the Sango Hills, near the mouth of the Kagera river; quartzite, slightly weathered. Repeated on p. 227.

Fig. 2. Another Proto-Tumbian pick, from the same locality as No. 1; quartzite, fresh.

Approx. $\frac{1}{2}$ scale.

I

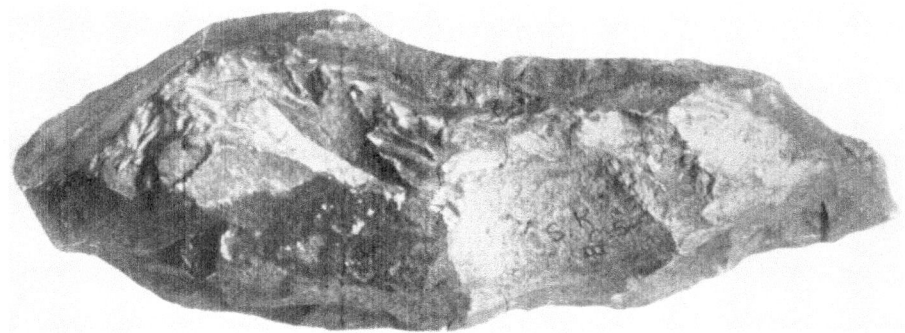

2

PLATE XVII

Tumbian Culture

Fig. 1. Two views of a Proto-Tumbian pick from the N-Horizon rubble, Kagera 100 ft. terrace, near Nsongezi; quartzite, a little weathered. Repeated on p. 229 (2).

Fig. 2. Side view of a flat-based Proto-Tumbian pick from the same locus; quartzite, slightly weathered.

Figs. 3 and 4. Small Proto-Tumbian hand-axe tools; same provenance, material and condition as above.

Fig. 5. Two views of a flat-bottomed, hump-backed Proto-Tumbian tool; same provenance, material and condition as above.

Fig. 6. A Proto-Tumbian pick. Same provenance, material and condition as above.

Approx. $\frac{1}{2}$ scale.

1 2

3 4

5 6

223

FIGURE 37

Tumbian Culture

A large Proto-Tumbian artifact of typical, so-called *Sangoan facies* from the Younger Rubble, Kagera valley. The tool has a pronounced base-keel towards which the two unworked under-surfaces converge. As is frequently the case with tools of this class, the two lower faces are natural joint or cleavage planes which have been utilised as striking platforms when the upper surface was being flaked. Few other specimens could show this technique to better advantage than this tool; quartzite, very slightly weathered.

$\frac{1}{2}$ scale.

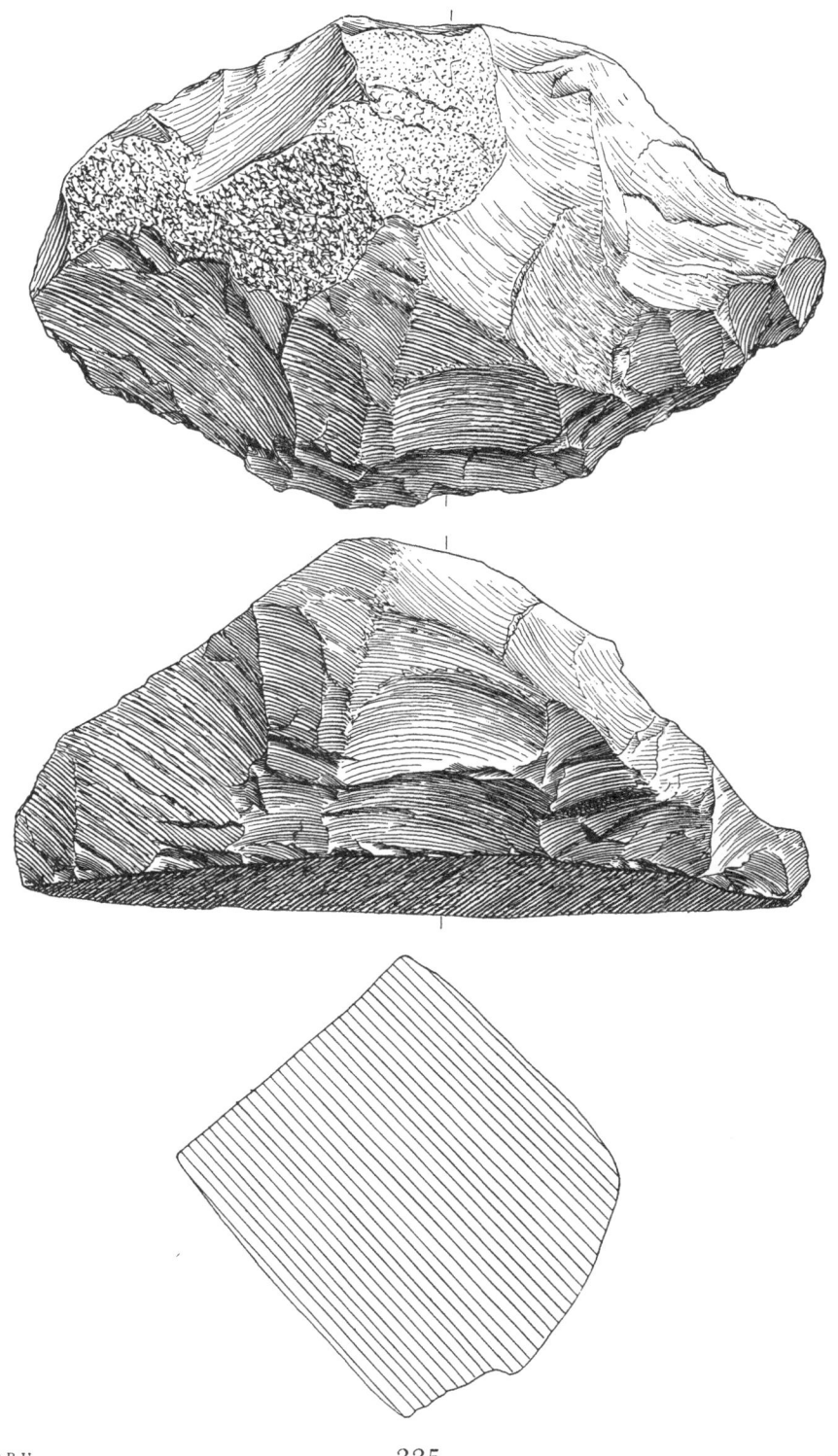

FIGURE 38

Tumbian Culture

Two views of a typical Proto-Tumbian pick from a rubble on one of the Sango Hills, near Lake Victoria; quartzite, slightly weathered.

½ scale.

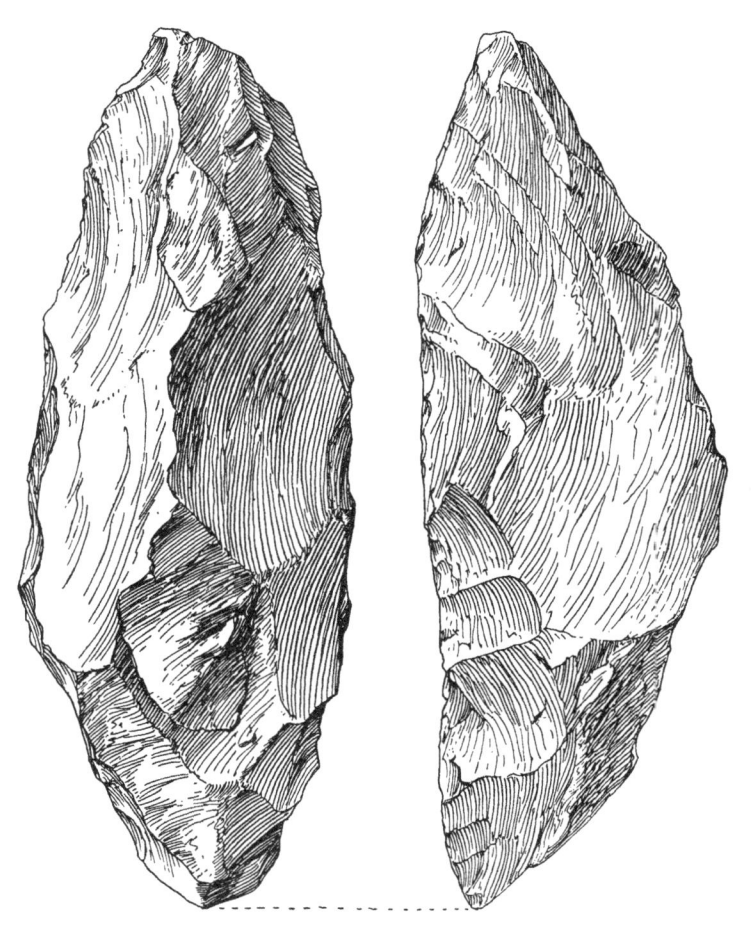

15·2

FIGURE 39

Tumbian Culture

Two Proto-Tumbian picks characteristic of the N-Horizon rubble, Kagera 100 ft. ±
terrace. The lower example was found *in situ*, but the upper tool was found in an
erosional basin cut through this horizon; quartzite, both slightly weathered.

½ scale.

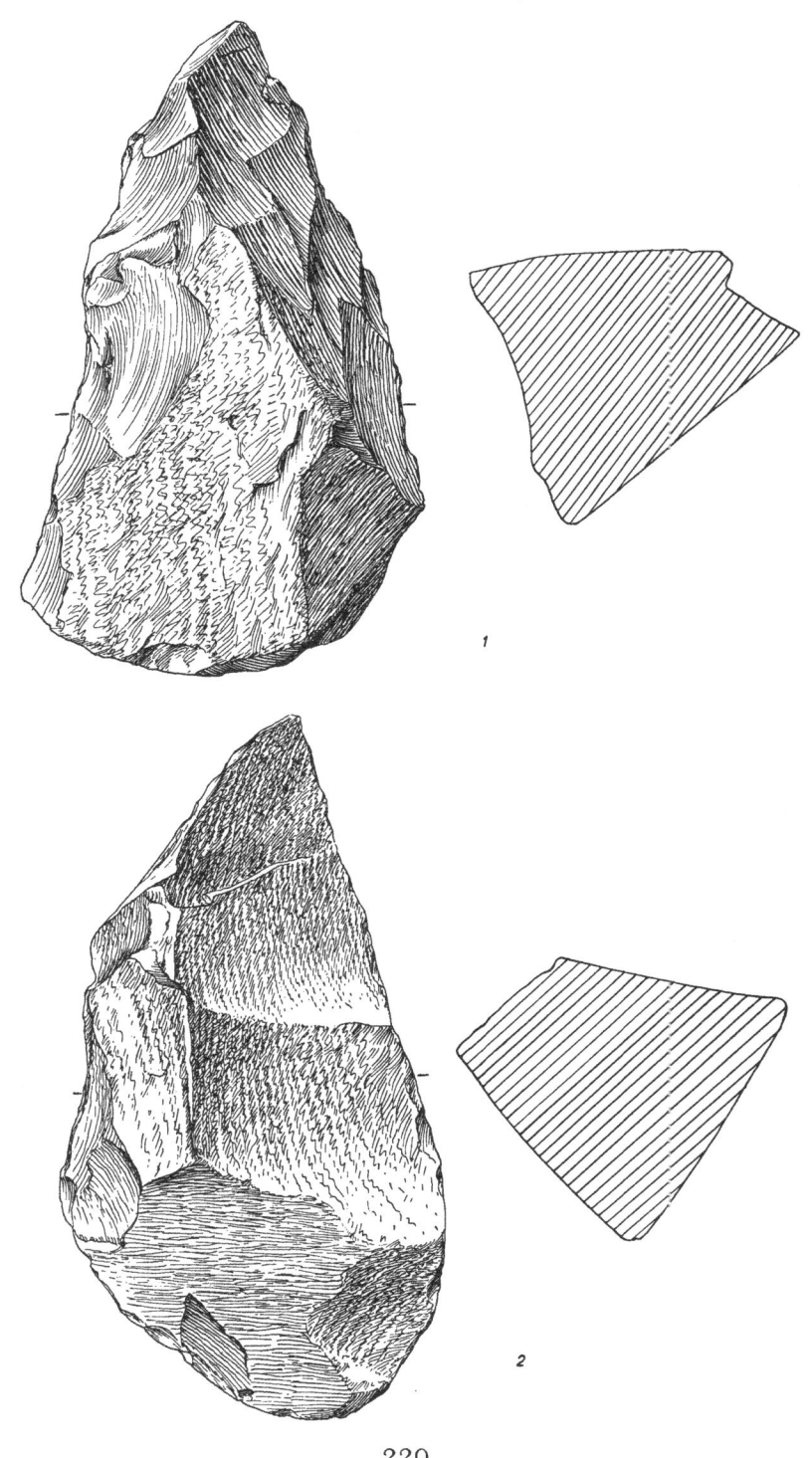

1

2

229

PLATE XVIII

Tumbian Culture

Fig. 1. Two views of a long Middle Tumbian pick-hand-axe made on a flake. From a washout cut into the upper clays of the Kagera 100 ft. ± terrace, near Nsongezi; quartzite, very slightly rolled on the edges.

Fig. 2. Two views of another Middle Tumbian pick-hand-axe found *in situ* in the upper 100 ft. ± terrace clays, near Nsongezi; quartzite, very slightly rolled on the edges. Repeated on p. 245 (2).

Approx. $\frac{1}{2}$ scale.

PLATE XIX

Tumbian Culture

Figs. 1 and 3. Two views, respectively, of two Middle Tumbian "duck-head" hand-axes from washouts in the upper 100 ft. ± terrace clays, near Nsongezi. Note the characteristic flattened underside (usually a plain flake surface) and thinning of the tip from the thick butt; quartzite, No. 1 slightly rolled, No. 3 fresh.

Fig. 2. A Middle Tumbian pick. Same provenance as Nos. 1 and 3; quartzite, fresh. Repeated on p. 257 (2).

Fig. 4. A Middle Tumbian "duck-head" hand-axe found *in situ* in the upper 100 ft. ± terrace clays, near Nsongezi; quartzite, fresh. Repeated on p. 245 (1).

Approx. $\frac{1}{2}$ scale.

1

2

3

4

233

PLATE XX

Tumbian Culture

Three views of a Middle Tumbian composite artifact—a sort of combined hand-axe and steep-scraper. Fig. *a* shows the tool resting partly on the right, hand-axe edge and partly on the flat, steep-scraper base. Fig. *b* shows the hand-axe in full face and the right edge of the steep-scraper, whose flat base now faces to the right. Fig. *c* shows the tool resting on the left, hand-axe edge and showing the flat underside of the steep-scraper. Found in a washout in the upper 100 ft. ± terrace clays, near Nsongezi; quartzite, fresh. Repeated on p. 249.

Approx. $\frac{1}{2}$ scale.

PLATE XXI

Tumbian Culture

Middle Tumbian flat *bifaces* from washouts in the upper 100 ft. ± terrace clays, near Nsongezi; quartzite, fresh. No. 1 repeated on p. 253 (1).

Approx. ½ scale.

1

2

3

4

237

PLATE XXII

Tumbian Culture

Fig. 1. Three aspects of a long Middle Tumbian lance head from washout in upper 100 ft. ± terrace clays, near Nsongezi; quartzite, very slightly weathered. Repeated on p. 251 (3).

Fig. 2. Two aspects of another, similar tool from the same locus; quartzite, fresh. Repeated on p. 251 (2).

Approx. ½ scale.

2

I

239

PLATE XXIII

Tumbian Culture

Fig. 1. Long Middle Tumbian lance head; quartzite, fresh. Repeated on p. 251 (1).

Fig. 2. Long Middle Tumbian leaf-shaped *biface*; grey quartz, fresh.

Fig. 3. Middle Tumbian cleaver; quartzite, fresh.

Fig. 4. Two aspects of a large Middle Tumbian tranchet; quartzite, fresh.

Fig. 5. Another large Middle Tumbian tranchet; quartzite, fresh. Repeated on p. 247 (1).

All these specimens are from washouts in the upper 100 ft. ± terrace clays, near Nsongezi.

Approx. $\frac{1}{2}$ scale.

1

2

3

4

5

PLATE XXIV

Tumbian Culture

Fig. 1. Large, flat Middle Tumbian *biface* made on a flake; quartzite, very slightly rolled.

Fig. 2. Two views of a Middle Tumbian cleaving tool on an end-blow flake; quartzite, fresh.

Both tools are from washouts in the upper 100 ft. $\pm$ terrace clays, near Nsongezi.

Approx. $\frac{1}{2}$ scale.

242

2

1

16-2

FIGURE 40

Tumbian Culture

Two Middle Tumbian hand-axes found *in situ* in the upper Kagera 100 ft. ± terrace clays, near Nsongezi. Note the tendency to the "duck-head" shape which is a special feature of Middle Tumbian hand-axes, and is probably derived from the hump-backed shapes in the Proto-Tumbian; quartzite, No. 1 fresh, 2 slightly rolled.

½ scale.

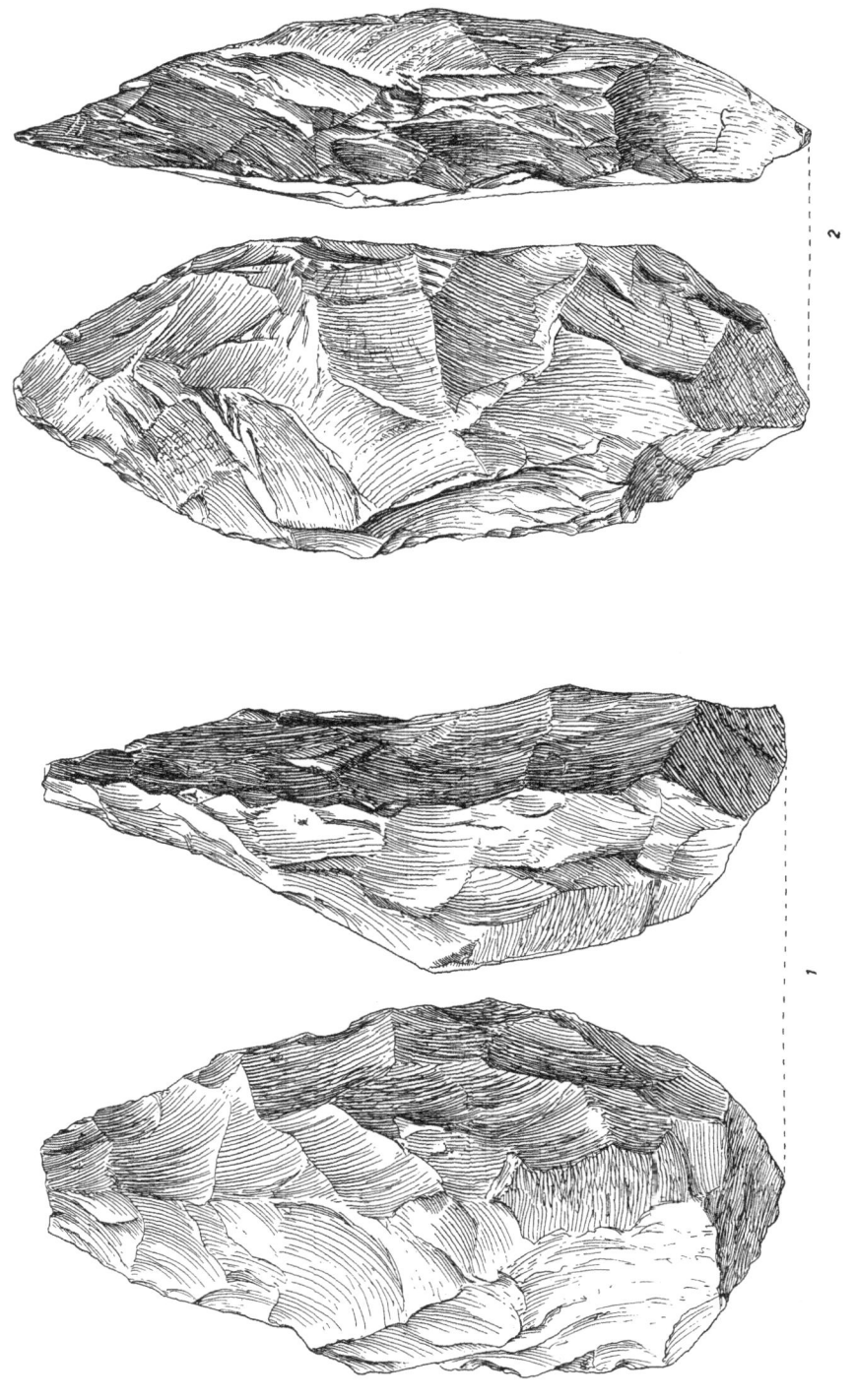

FIGURE 41

Tumbian Culture

Two Middle Tumbian implements from washouts in the upper clay, Kagera 100 ft. ±
terrace, near Nsongezi. No. 1 is the largest example of the Tumbian type of tranchet
found by us, nor do I know of a larger in any of the museum collections of Tumbian
material from the French and Belgian Congo regions. No. 2 is an example of the
fairly common Middle Tumbian combination tool, a hand-axe-scraper; both quartzite, fresh.

$\frac{1}{2}$ scale.

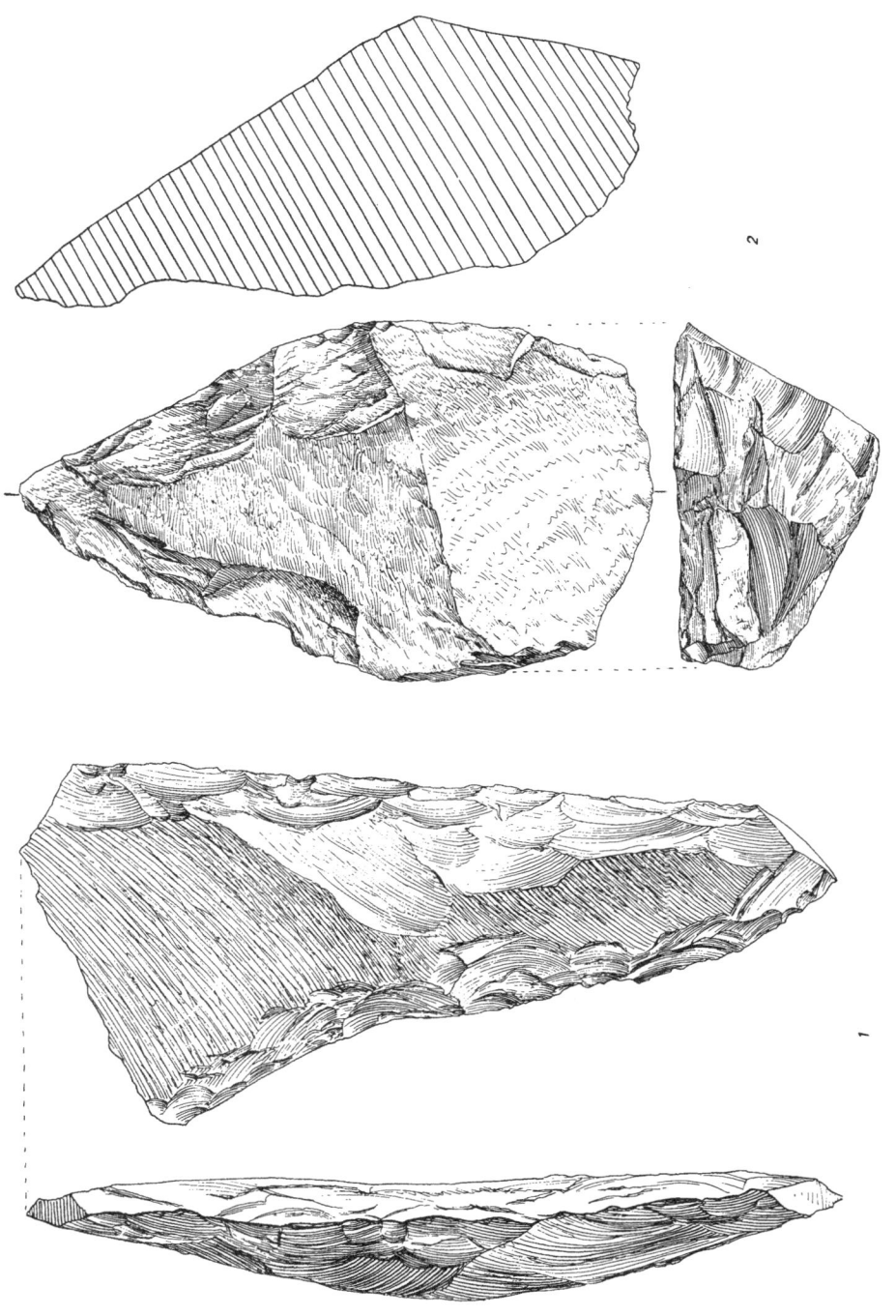

FIGURE 42

Tumbian Culture

Four aspects of another interesting Middle Tumbian double-utility implement from a washout in the upper clays, Kagera 100 ft. ± terrace, near Nsongezi. *a* and *b* show the upper and lower major aspects, respectively. *c* shows the tool resting on the flat, steep-scraper base, while *d* shows the scraper base itself which is a natural surface; quartzite, fresh.

$\frac{1}{2}$ scale.

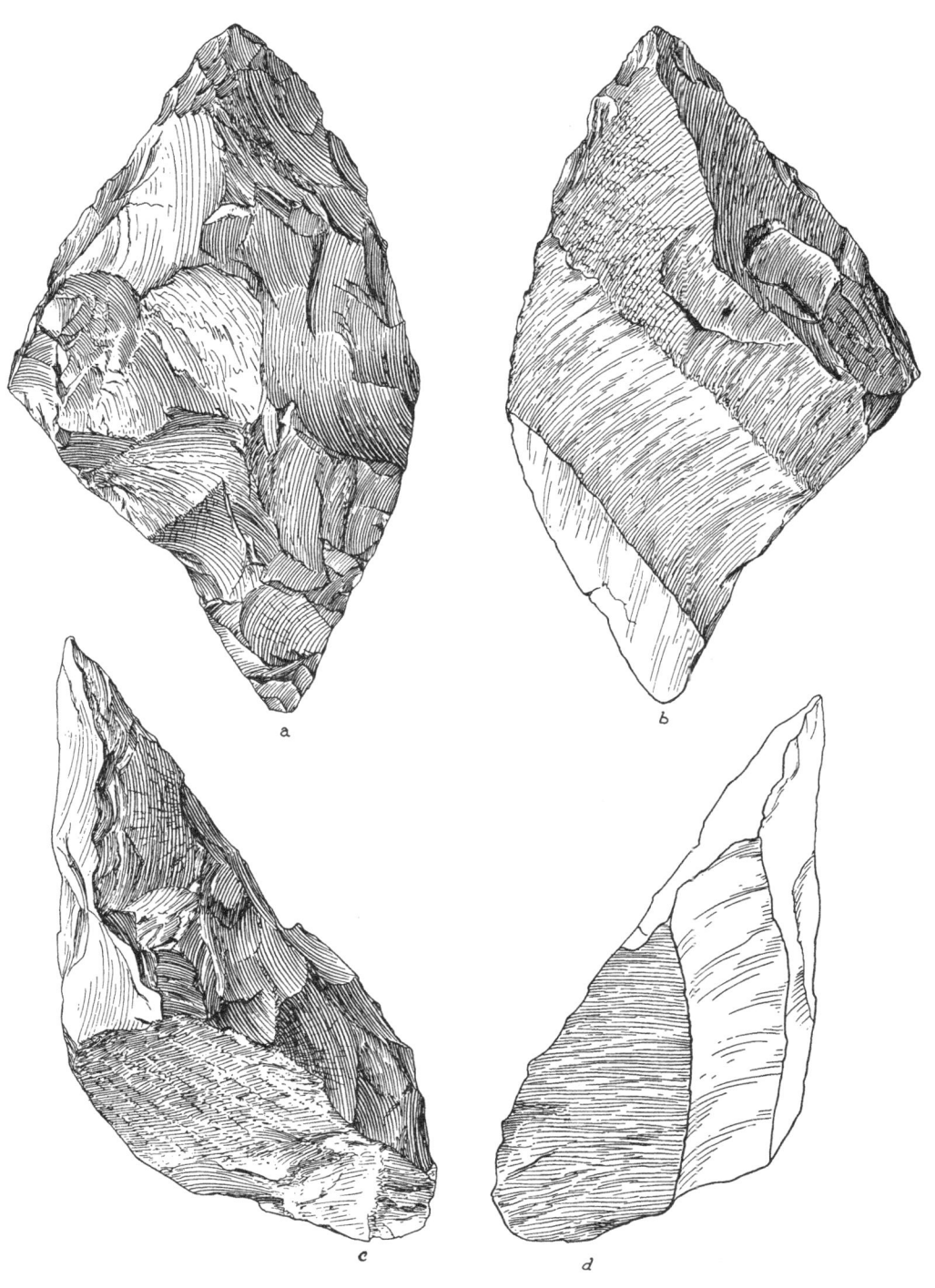

a

b

c

d

249

FIGURE 43

Tumbian Culture

Three long Middle Tumbian leaf-shaped *bifaces* from washouts in the upper clays, Kagera 100 ft. ± terrace, near Nsongezi. Nos. 1 and 2 are made in a dark grey quartzite and are fresh, while No. 3 is pale brown quartzite, very slightly weathered.

½ scale.

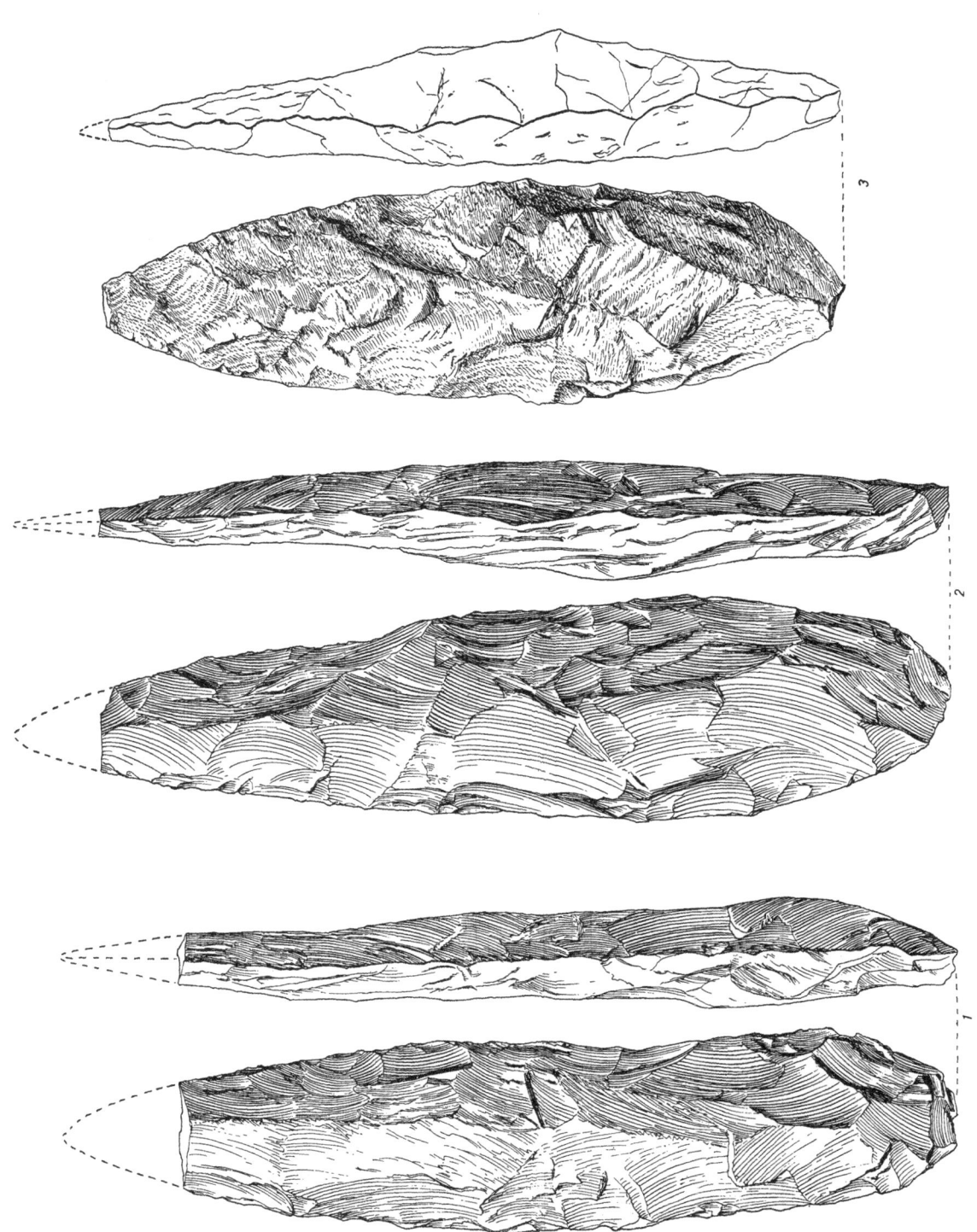

FIGURE 44

Tumbian Culture

Five Middle Tumbian *bifaces*. All except No. 4 are from washouts in the upper clays, Kagera 100 ft. ± terrace, near Nsongezi. No. 4 was found *in situ* in these clays by Mr E. J. Wayland, whom I thank for permission to figure this valuable specimen; it is made in a rather cherty quartzite and is very slightly worn; the rest are in quartzite, fresh.

$\frac{1}{2}$ scale.

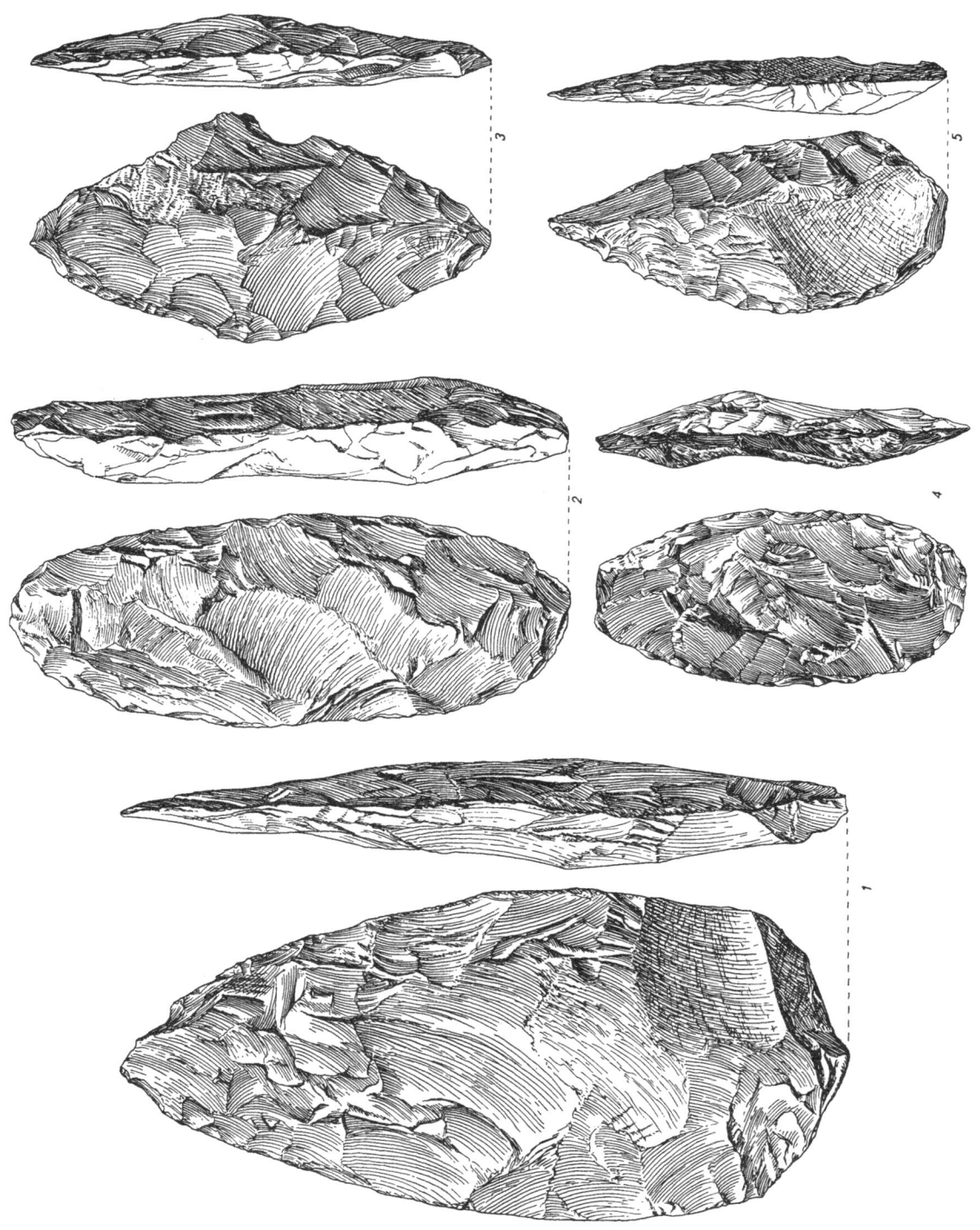

FIGURE 45

Tumbian Culture

Middle Tumbian tools from washouts in the upper clays, Kagera 100 ft. ± terrace, near Nsongezi. No. 1 is a beautiful specimen (broken at the butt) in pure white quartz. Nos. 2, 3, 6 and 7 are variously shaped points. Nos. 4, 5, 9 and 10 are tranchets. No. 8 is a rare example of the Middle Tumbian cleaver type; all except Nos. 1 and 3 (white quartz) are in dark grey quartzite, fresh.

$\frac{1}{2}$ scale.

254

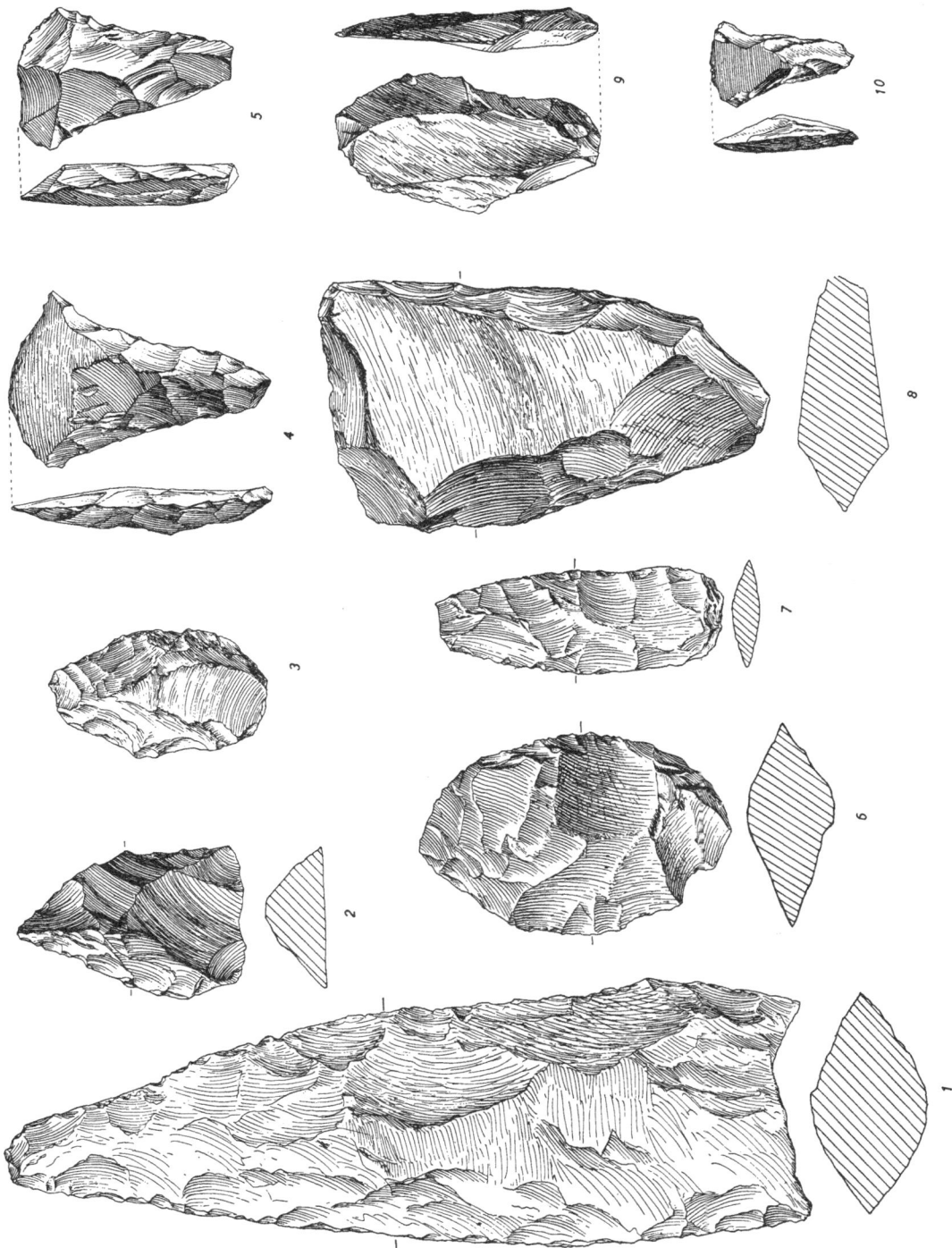

255

FIGURE 46

Tumbian Culture

Middle Tumbian tools from washouts in the upper clays, Kagera 100 ft. $\pm$ terrace, near Nsongezi. Nos. 1 and 2 are examples of Middle Tumbian picks, No. 3 is a highly interesting long-flake core, round in section and flat-based. The rest are end-scrapers; quartzite, fresh.

$\frac{1}{2}$ scale.

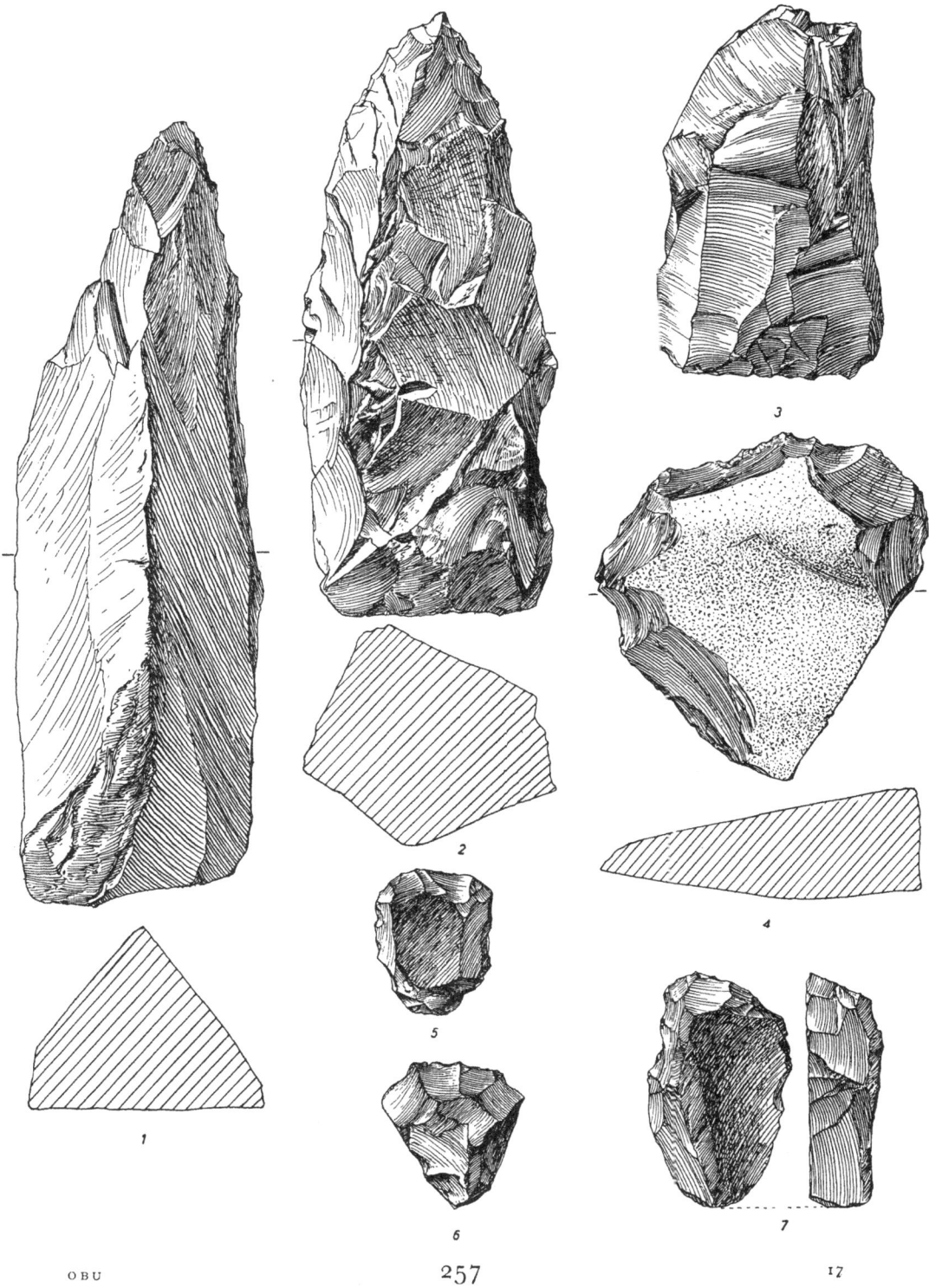

FIGURE 47

Tumbian Culture

Late Tumbian implements (perhaps of Mesolithic age) found in washouts near Nsongezi. Their position, high up the slopes of these erosional basins, suggested that they were washed out from between the top of the Kagera 100 ft. $\pm$ terrace beds and the surface mantle of brown earth. Their Mesolithic age was suggested when one tranchet of similar type to those in the drawings was found in the Magosian levels at the Kagade rock-shelter; white quartz, fresh.

Nos. 1–6: Points.

Nos. 7–13: Tranchets.

No. 14: Crescent.

No. 15: Scraper.

No. 16: Small core.

$\frac{1}{2}$ scale.

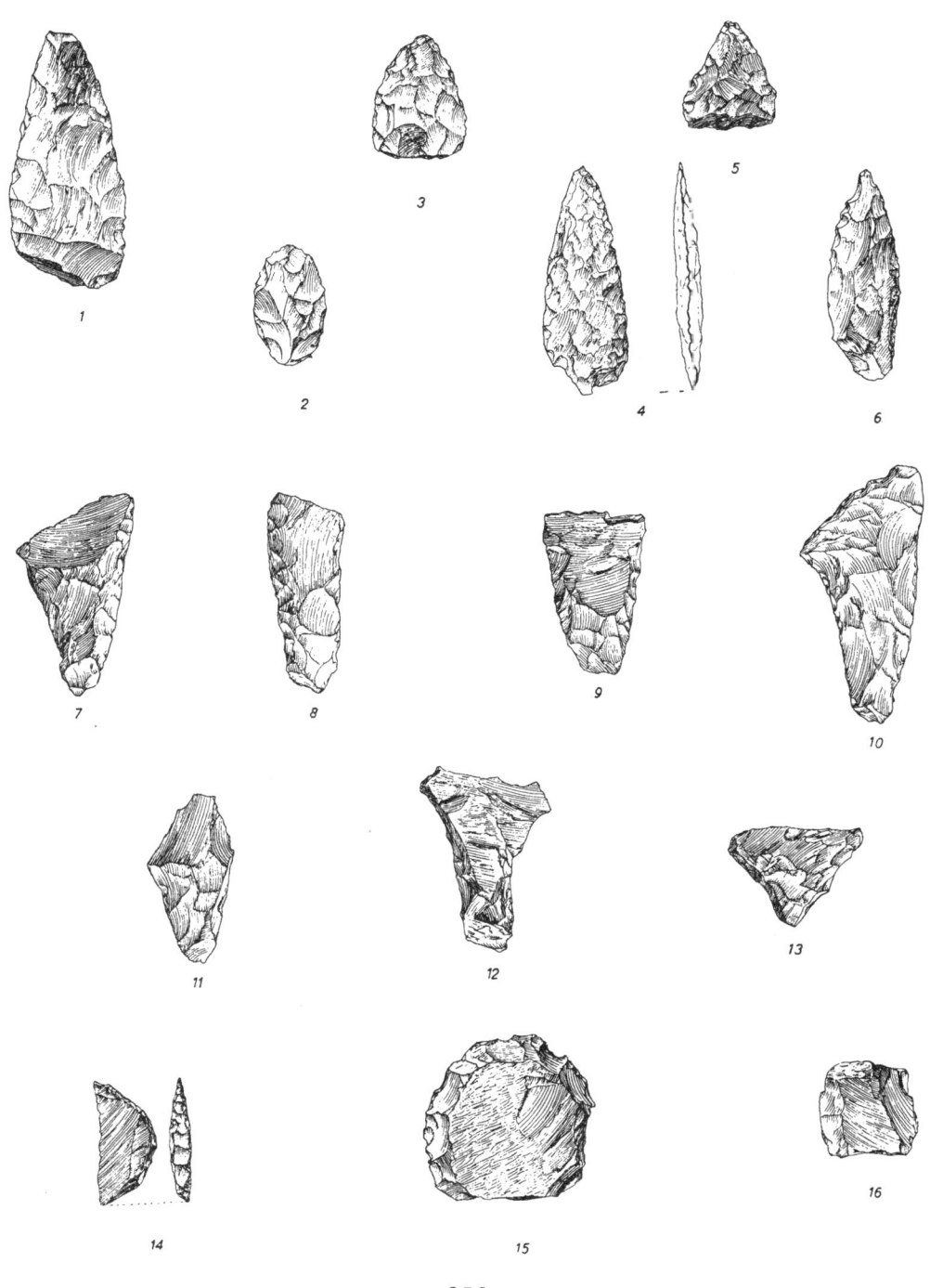

CHAPTER XII

Uganda Mesolithic

MAGOSIAN

UNTIL the description of this industry by Wayland and Burkitt in 1932,[1] research into the East African Levalloisian and its derivatives had suggested that these terminated in the final Still Bay which, in Kenya, last appeared in the Upper Gamblian. They seemed to have died out completely by the time the Elmenteitan—a Kenya Capsian derivative—appeared in the Makalian period, for there was no trace of Still Bay influence in this industry nor in the somewhat later Wilton A, which is considered by Leakey to be an offshoot of his Upper Kenya Aurignacian.

The silted-up water-hole, or granite cistern site at Magosi, in the extreme east of Uganda was of special value, therefore, since it provided evidence not only of the late survival of the Still Bay, but also of its connection with the Wilton culture. The material was studied by Burkitt and, on account of the association of some Still Bay tools with others of crude Wilton facies, he suggested tentatively that the Magosian was a form of proto-Wilton. He did, however, also recognise the possibility that the industry might have been the result of contact between a late Still Bay stage and an Early Wilton, already in existence —a possibility which is, to my mind, the more likely.

Since the Magosian of the type station has been described, it has become clear that it had a considerable, if sporadic, range in East Africa. As long ago as 1901, the Vicomte de Bourg de Bozas made an extensive collection of stone tools, during an exploratory visit to Abyssinia, and much of this material has turned out to be Mousterian, Still Bay, Magosian and Wilton.[2] It is claimed that there is a continuous development from the earliest to the latest of these, but, as the material is all from the surface, and could only be divided by study of the relative

[1] The Magosian Culture of Uganda, *J.R.A.I.* vol. LXII, July–December, 1932.
[2] The Abbé Breuil and Harper Kelley, *Journ. de la Société des Africanistes*, tome VI, 1936, pp. 111 *et seqq.*

states of preservation, this evolution is not absolutely certain. Furthermore, judging solely by the specimens illustrated, it looks as if most of the material described as Magosian is actually an admixture of Still Bay and true Wilton of the same age. It does not give the impression of being ancestral to the younger industry described as Wilton.

In 1932, Leakey excavated a rock-shelter at Apis Rock, Tanganyika, the results of which were briefly described recently.[1] He describes how the Still Bay of the lower part of the section was followed immediately by a well-developed Magosian industry and states that the latter was undoubtedly derived from the former. He admits, however, that the Apis Rock Magosian is most probably a later stage than that at Magosi itself, as it is associated with pottery. His illustrations seem to bear out this suggestion, for the tools have every appearance of being later than the series at Magosi, being, on the whole, much better made. If this is so, the Apis Rock Still Bay cannot be ancestral to the local Magosian unless this industry was born at different times in different places and, moreover, in varied states of maturity at various sites.

The date of the Magosian industry at the type station was tentatively suggested by Wayland as being inter-Gamblian-Makalian. This conclusion was reached by a logical train of argument, taking into consideration the nature and position of the site—a pre-Magosian, water-excavated rock cistern in a semi-desert country—and applying the previously accepted ideas concerning East African Upper Palaeolithic climate and contemporary industries. When Wayland wrote, however, it was not known that the Late Still Bay occurred as late as this same inter-Gamblian-Makalian dry period, while it is far from certain to-day either that the Gamblian period was very noticeably wet or that the interval between it and the Makalian was of much significance as a dry epoch. Thus, there is no means of knowing, at present, if the Still Bay element was or was not of Makalian age, when it could have come into contact with a true Wilton, albeit of an earlier stage than that appearing near the end of the Makalian in Kenya.

Leakey's argument that, at Apis Rock, the conditions were such that man could not have lived there during the Gamblian-Makalian dry period, is not supported by any evidence of the degree of aridity

[1] Leakey, *Stone Age Africa*, 1936.

prevailing at the time and is further contradicted by his own statement that the Still Bay industry there was "followed immediately" by the Magosian which was derived from it. He says, further, that the Apis Rock Magosian is an Upper stage of Makalian age, but does not explain how there could have been unbroken occupation of the shelter from Lower Still Bay to Upper Magosian times right through the period supposedly too dry for any habitation at all.

In the absence of definite proof one way or the other, which will almost certainly be forthcoming if and when the Magosian is discovered in the lake deposits in Kenya, the present writer can only record what seems to him to be the most likely origin of this culture, that is, that it represents a contact, at some time during the Makalian period, between a very late stage of Still Bay, probably long influenced by an earlier contact with one or other of the Kenya Capsian-like industries, and an early stage of Wilton, already in existence. In this connection, it is worth noting that Leakey himself regards his Wilton A (without Still Bay elements) as an offshoot of his Late Kenya Aurignacian.

Technique and Typology

The implementiferous deposit at Magosi was 11 ft. deep and lay against the sloping side of the rock-cistern which rises in a series of steps from bottom to top.

The tools show little change throughout the deposit, though those at the very bottom are a little more primitive than the rest and those at the very top contain types not found in the other levels and probably indicate later admixture.

Two groups, representing the two probable lines of ancestry, Wilton and Still Bay, are present in all levels below the top, one consisting of pygmy types, with cores, lunates, backed-blades, awls, thumb-nail scrapers and burins, and the other including points, trimmed on one or both faces, discs of various sizes and cores resembling the "tortoise" type.

The materials employed include chert, quartz, chalcedony and obsidian.

Uganda Mesolithic

KAGADE MAGOSIAN

The only industry of Magosian type that we ourselves found in Uganda occurred at a granite rock-shelter at Kagade in the west of the Mubende district, not many miles from the edge of the Albert Rift scarp. The deposit was 6 ft. 6 in. in depth.

The industry is very like that from the type station, except that it is altogether poorer and there are no burins. The tools comprise points, lunates, end- and side-scrapers, small discoid cores, conical cores and hammer-stones, in white quartz. Red ochre and pieces of used haematite occurred at all levels. There was no pottery. One small tranchet of Late Tumbian type was found and appeared to be an importation into the site, as surface finds of similar tools were fairly numerous in the neighbourhood.

FIGURE 48

Mesolithic

Magosian tools from a granite rock-shelter at Kagade. No. 1 is actually a Late Tumbian tranchet which must have been brought into the shelter by the Magosian occupants. Similar tranchets and typical Late Tumbian points are fairly common surface finds in this neighbourhood. The material of the following specimens is white quartz and the tools are fresh.

No. 1: Tranchet; 1 ft. 6 in. level.

No. 2: Scraper-awl; 5 ft. 6 in. to 6 ft. level.

Nos. 3 and 4: Small Magosian points derived from the Late Still Bay; 6 ft. to 6 ft. 6 in. level.

No. 5: Awl on flake with faceted butt; 6 ft. to 6 ft. 6 in. level.

No. 6: Small tortoise core; 6 ft. to 6 ft. 6 in. level.

No. 7: Small discoid core; 6 ft. to 6 ft. 6 in. level.

$\frac{1}{1}$ scale.

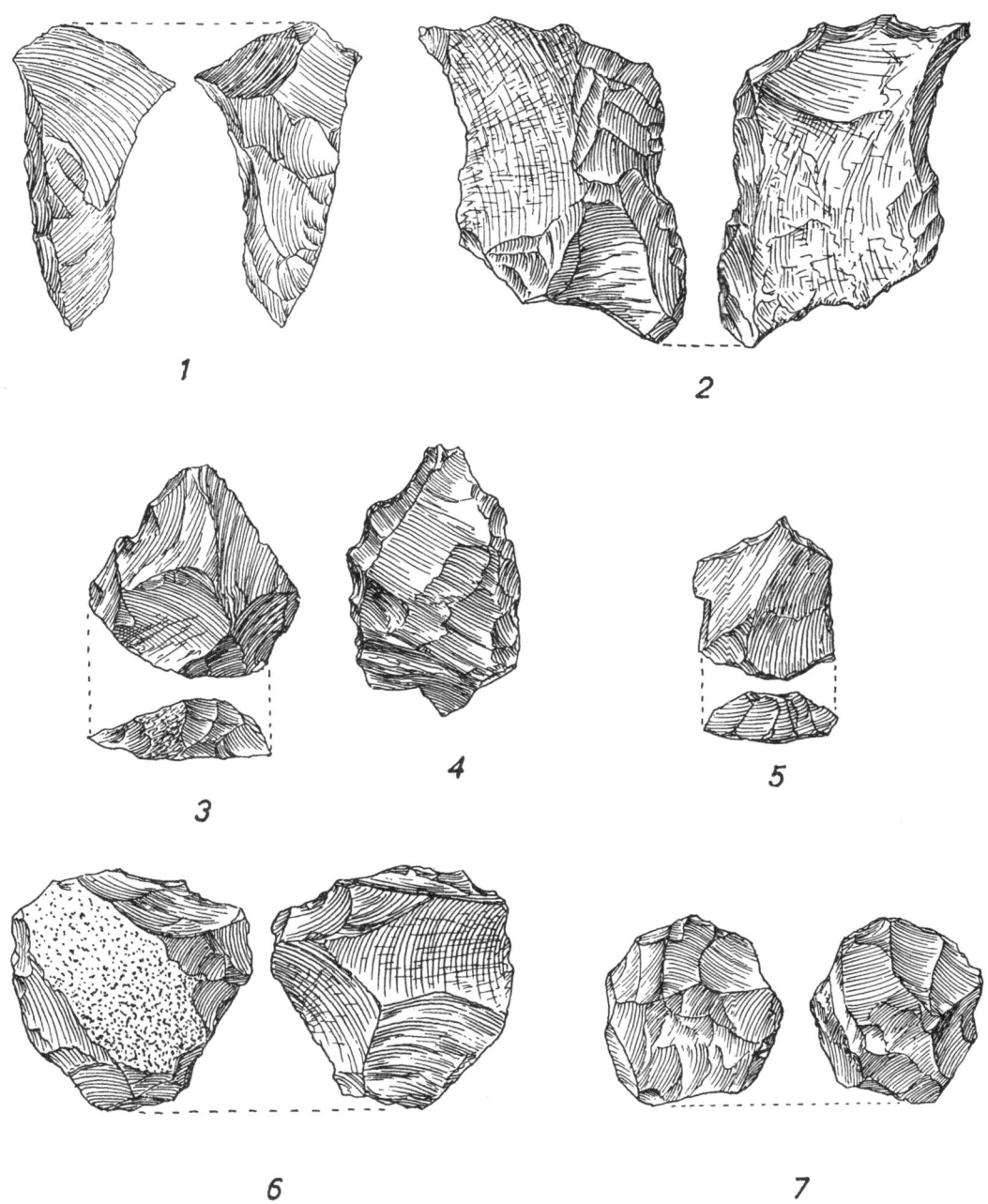

1 2

3 4 5

6 7

FIGURE 49

Mesolithic

Magosian tools from a granite rock-shelter at Kagade; white quartz, fresh.

Nos. 1 and 2: Points; from 1 ft. 6 in. level.

No. 3: Small microlithic core; 1 ft. 6 in. level.

No. 4: Hammer stone; 1 ft. 6 in. level.

No. 5: Piece of worn haematite; 1 ft. 6 in. level.

No. 6: Discoid core; 1 ft. 6 in. level.

No. 7: Small tortoise core; 1 ft. 6 in. level.

No. 8: End-scraper; 1 ft. 6 in. level.

No. 9: Point; 1 ft. 6 in. level.

Nos. 10, 11 and 12: Points; 2 ft. to 2 ft. 6 in. level.

No. 13: Narrow object with blunting on one side; 2 ft. to 2 ft. 6 in. level.

Nos. 14 and 15: Flakes with rough blunting on thick edges; 2 ft. to 2 ft. 6 in. level.

$\frac{1}{1}$ scale.

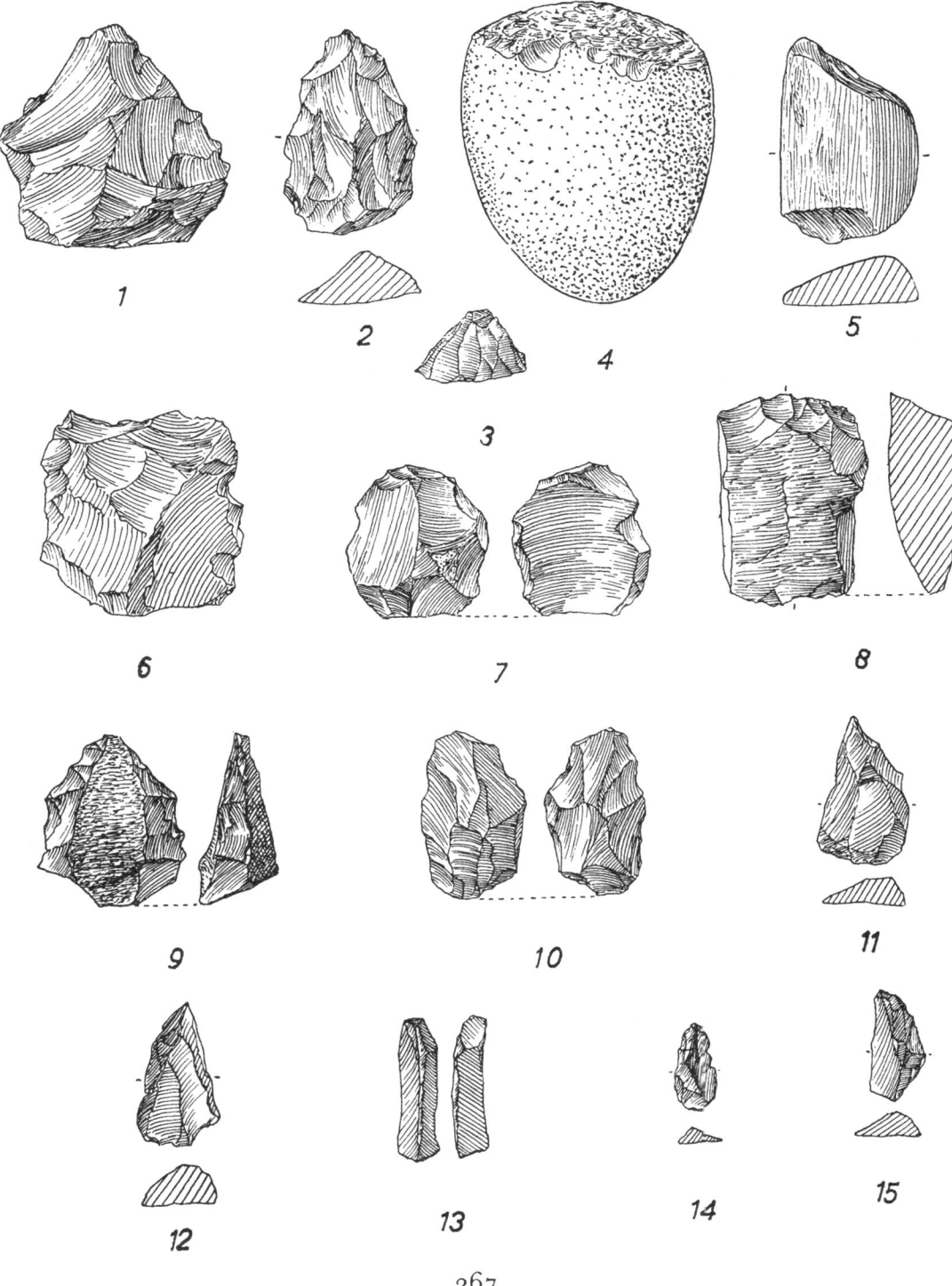

1

2

3

4

5

6

7

8

9

10

11

12

13

14

15

267

CHAPTER XIII

Uganda Neolithic Industries

WE know of no polished tools in Uganda and the following industries are all of microlithic type except one, the Kageran. As might be expected, they bear some resemblance to the Late Stone Age industries of Kenya, though the Wilton element appears to predominate.

Owing to the lack of geological data and the fact that the fauna is always a modern one, it is difficult to date these industries exactly, either individually or with reference to one another. All that we can say is that, at Nsongezi, the Kageran is followed by the Wilton-Neolithic A and that, at the Walasi open stations, the Still Bay is followed by a late, Magosian-like industry which, in turn, is succeeded by the Wilton-Neolithic B.

KAGERAN

This industry, which I have named after the Kagera River, on whose banks it was found, was first discovered by Wayland, at a site on the 30 ft. ± terrace, just below the P.W.D. Camp at Nsongezi and again, poorly represented, in a rock shelter about a hundred yards away. In each case, it immediately preceded a microlithic industry, to be described in the next section.

The Kageran in the shelter lay at the base of the deposit, while, at the open station, it occurred in a definite horizon on the disturbed shingle, mixed with black earth, of the 30 ft. terrace, under 8 in. of black surface soil. At this site, two or three of the later, microlithic tools were also found in the 8 in. level and a number of them in the shingle below it, but this is due, I think, to the tiny implements having worked their way down, as they may easily do in such a soft, loose medium.

The Kageran is an unspecialised industry, made of blue quartzite, often on pebbles or small boulders that were to be found mixed with the shingle of the 30 ft. terrace. The tools are fairly large and consist of cores, choppers, some scrapers made on chunks and numerous flakes.

One or two of the implements have been made on older, heavily rolled tools.

An extensive rock outcrop projects above the level of the 30 ft. terrace a few yards from this site, and on the top and one side of this rock are a number of irregularly spaced hollows, clearly of human origin. They can only have been bored by some hard material and I feel very inclined to regard them as the work of the Kageran people. Possibly, excavation round the base of the rock would reveal some of the very tools worn in boring the little cup-holes.

Wilton-Neolithic A

This is the name which I have given to the industry from a rock shelter at Nsongezi, first discovered and tested by Wayland. Similar material was found by him on the 30 ft. terrace near by and he sent a series from each site to Cambridge, where it was handed over to me for study. Neither he nor I knew of any parallel to the quartzite group, but I was inclined to regard the microlithic industry as a form of the Gumban found by Leakey in Kenya. Having excavated the rest of the shelter and a further site on the terrace, however, I think that the entire absence of polished tools, stone bowls and querns and of pottery in the lower levels precludes a close kinship with the Gumban, while the presence of thumb-nail scrapers seems to indicate a Wilton tradition, assuming these tools to be characteristic of this industry and its derivatives, hence the name Wilton-Neolithic A; Wilton-Neolithic B is described later.

The shelter is in the gorge of the Kagera valley, west of Nsongezi, and had been excavated by the river in a rubbly conglomerate that, at one time, filled the bottom of the valley. The shelter occupies a position on the left bank of the river, below the Nsongezi-Kikagati road and a little above the 30 ft. terrace. The valley here is clothed with patches of trees and thick bush, with intervals of thinner bush and grass.

The shelter is not very large and has no terrace to speak of, as the ground outside slopes steeply down to the 30 ft. level and again to the river. The deposit was about 8 ft. deep and contained many large rocks from the roof and sides. Wayland had removed the top 2 ft. of recent soil and stones from the whole area before beginning his test trench, so, in conformity with his levels, the first 2 ft. are disregarded in our work also. As usual, the shelter was longer from end to end than from front to back, and the outer edges, lying rather beyond the overhang of the roof, were less productive than the middle part, which contained the main hearths and where the deposit was thickest and very ashy.

Two industries were present, a scanty selection of Kageran at the bottom and the Wilton-Neolithic A above. A layer of *Etheria elliptica* shells occurred from 3 ft. to 3 ft. 6 in. and one of *Parreysia bakeri* from 4 ft. to 4 ft. 6 in., while calcined bones were common. According to Dr A. T. Hopwood,[1] the majority of the fauna consists of antelopes, and it is all modern.

It would appear that two stages of Wilton-Neolithic A occur in the shelter because, although the tool types are the same throughout the deposit, only varying in numbers, pottery is abundant in the top 4 ft., below which it ceases. This pottery is, on the whole, well made and decorated with various incised designs, though some very rough sherds were also present.

The first 2 ft. must be regarded as slightly disturbed, since two metal objects were found—an iron arrow-head and spear-butt—while the levels from 4 to 6 ft. were the most productive, tools decreasing from 6 ft. to the bottom which was almost barren.

The implements are made of white quartz and rock-crystal, natural crystals of the latter being, apparently, attractive to these people, as quite a number were found, sometimes worked into little awls or points. There were also lunates, backed blades, segmental pebble knives, scrapers, chisels, points, cores and flakes and prodigious quantities of chips. One polisher and a round bead of ostrich egg-shell and a square one of oyster shell were also found.

The lunates were of two types, the usual crescent, with a straight blade and curved back, artificially blunted, and another, whose back was formed by the natural outer surface of the quartz pebble from which the flake had been struck, thus requiring no further blunting. Larger and more asymmetrical tools of this class have been referred to as segmental pebble knives.

[1] See Appendix B, The Mammalian Fossils.

The backed blades were very small and finely worked to a sharp, tapering point.

There were a few end- and side-scrapers and some thumb-nail scrapers of Wilton type, though none of these tools were very numerous.

The cores included discoid, bipolar and conical types and were very abundant, as were the flakes.

It is not possible to date either the Kageran or the Wilton-Neolithic with any accuracy at present, since neither is exactly comparable to any other industry known and dated in East Africa. From its typology, the Wilton-Neolithic A is probably a late form or descendant of the East African Wilton, of Neolithic date and the Kageran apparently precedes it slightly.

WILTON-NEOLITHIC B

The site of this industry was the Chui (Leopard) Cave, at the foot of towering cliffs at the north-west end of Walasi Hill, Mount Elgon. There was a little recent Bantu material on the surface of the floor of the cave and in the terrace beyond its mouth, but the main deposit was almost pure wood-ash and contained stone implements and bones, but no pottery.

The industry is an impoverished, microlithic group, but is of some interest, in spite of its rather miserable appearance, since, in addition to small cores and flakes and occasional lunates and Wiltonesque scrapers, several rough, but unmistakable burins occurred, almost the only representatives of this class of tool that we found in Uganda. Also, apart from the microliths, there were a number of larger stones showing various signs of use; one was a big, shapeless lump, with three holes bored in it, suggesting a fire-stone;[1] several others were worn smooth on one surface, due to use as pestles or grinders, and a quantity of smooth, round pebbles had been imported, probably from the Siroko River, about 3 miles away, most likely for sling-stones.

The presence of these used stones and the burins and the absence of backed blades and pottery show that this group is a distinct variation of the industry from Nsongezi. It approaches more nearly to the Gumban B of Kenya; but, here again, the absence of pottery and true stone

---

[1] For protecting the hand when twirling one stick against another.

271

bowls and querns and the presence of burins in the Walasi group, differentiates the two to a large extent. Assuming, as before, that the thumb-nail scraper is the distinguishing mark of the Wilton and its derivatives, the Walasi assemblage may be classed as Wilton-Neolithic B. This does not, necessarily, imply that it is later in time than Group A, for there is, at present, no more evidence than has been stated here by which either group may be dated.

The fauna, as identified by Dr A. T. Hopwood, is quite modern and includes the existing African elephant. An enormous quantity of charred rodent bones was present almost throughout the deposit.

OTHER NEOLITHIC REMAINS

Under this heading may be placed a short description of two quite isolated finds of curious spherical pots that are probably of Neolithic date.

The first group of pots was discovered while digging at the Bugungu Railway Pit site (see p. 291). Three almost spherical vessels lay side by side, mouths upwards, on a rough clay "dish" or platform, under about 5 ft. of stoneless brickearth and just above the Levalloisian rubble. All three were in an extremely friable condition, crumbling almost at a touch. Harder pieces show in section a typical red, oxidised band indicating definite burning and not merely sun-drying. Impressions of chopped grass are also visible.

The whole outer surfaces of the pots have a pecked appearance due to minute jab-marks made with a pointed object. There was no sign of smoke blackening.

None of the vessels had any form of rim lip to the small circular mouth which was thus merely a hole in the circumference, but each vessel was slightly flattened on the surface opposite the mouth.

When found the pots were naturally filled with earth and two of them appeared to have been used by millipedes as a cemetery. They were carefully emptied and searched for signs of former contents but contained no evidence of bones or of use as ritual offerings.

No stone tools accompanied the pots or were present in the brickearth at the same level, but a quantity of burnt earth and fragments of charcoal was found in the latter some distance away.

The second find almost exactly duplicated the first, although made on the banks of the Muzizi River, about 170 miles from Bugungu as the crow flies.

In this case, there were again three vessels and with them was found a single microlith to give a clue to their age. The vessels again lay upon a flat piece of clay, with mouths uppermost and about 3 ft. deep in a grey, earthy deposit just above flood plain level, near the road bridge.

In shape, size and appearance they are exactly similar to the Bugungu examples and also exhibit the same pecked technique of decoration.

The single artifact referred to is a small awl in white quartz. It was found within 6 in. of the pottery group. No other objects were associated with these finds and, as in the case of the Bugungu specimens, there was no evidence as to their use as ritual offerings.

FIGURE 50

Neolithic

Tools of the Kageran industry found at an open station on the Kagera 30 ft. ± terrace, at Nsongezi. No. 1 is a derived core-tool from the M-Horizon and was much rolled in the 30 ft. ± terrace gravels before being selected and rechipped slightly by the Kageran people. No. 2 is a chopper on a pebble, No. 3 is an axe or cleaver (note the shatter chips removed from the left edge); all made in quartzite, very fresh.

$\frac{2}{3}$ scale.

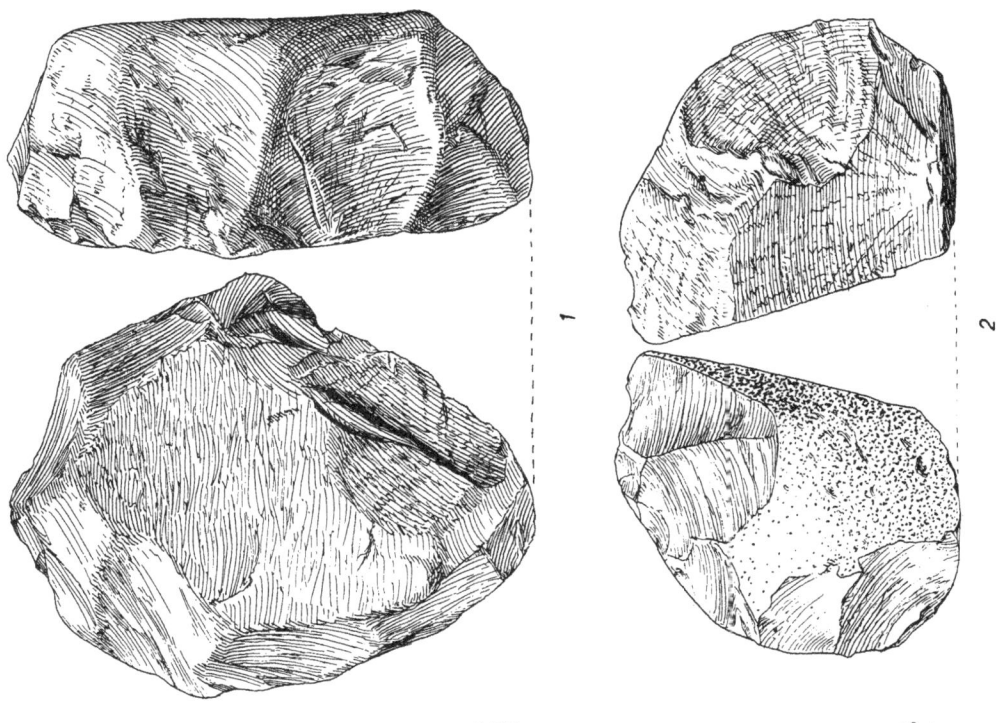

275

18-2

FIGURE 51

Neolithic

Further examples of Kageran implements from an open station on the Kagera 30 ft. ± terrace, at Nsongezi. No. 1 can only be described as a burin, Nos. 2 and 3 are scrapers and No. 4 is a small chopper on a pebble; quartzite, very fresh.

$\frac{2}{3}$ scale.

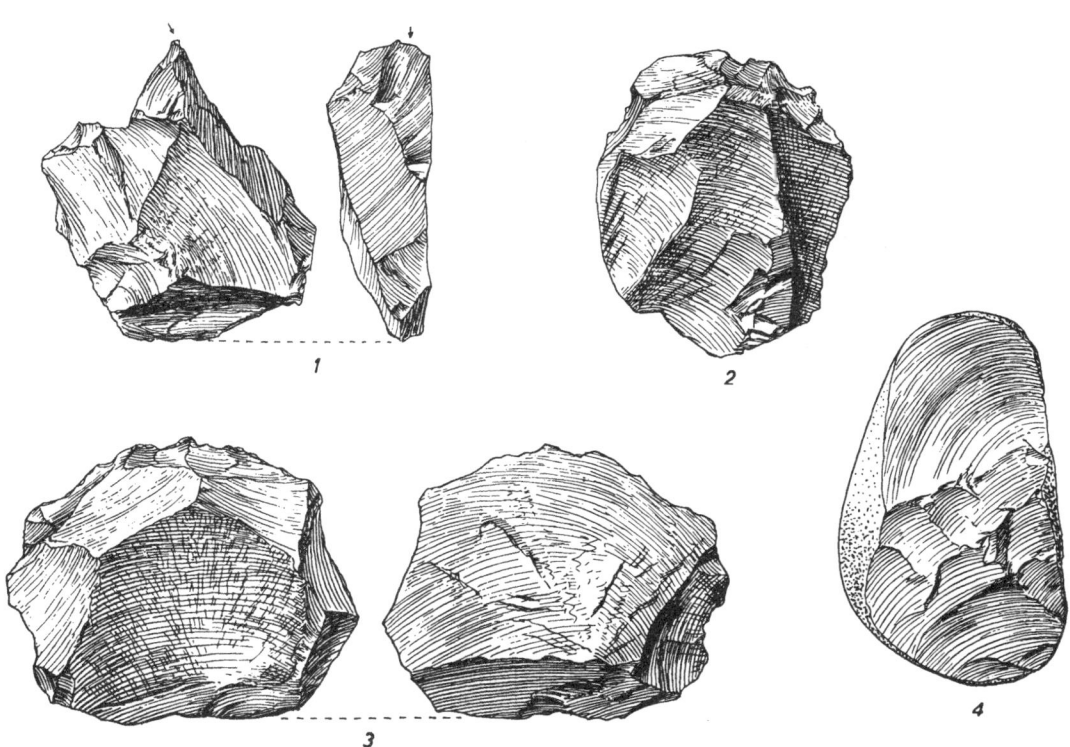

1   2   3   4

277

FIGURE 52

Neolithic

Specimens from the Wilton-Neolithic A industry at the rock-shelter at Nsongezi. The figured tools are from various levels, there being no observable difference in the industry from top to bottom except in the pottery (p. 285); white quartz and rock crystal, fresh.

Nos. 1–8: Crescents.

Nos. 9–13: Backed blades.

Nos. 14–16: Steep-sided cores.

No. 17: Rock crystal awl.

Nos. 18–19: Steep-sided cores.

No. 20: Discoid core.

$\frac{1}{1}$ scale.

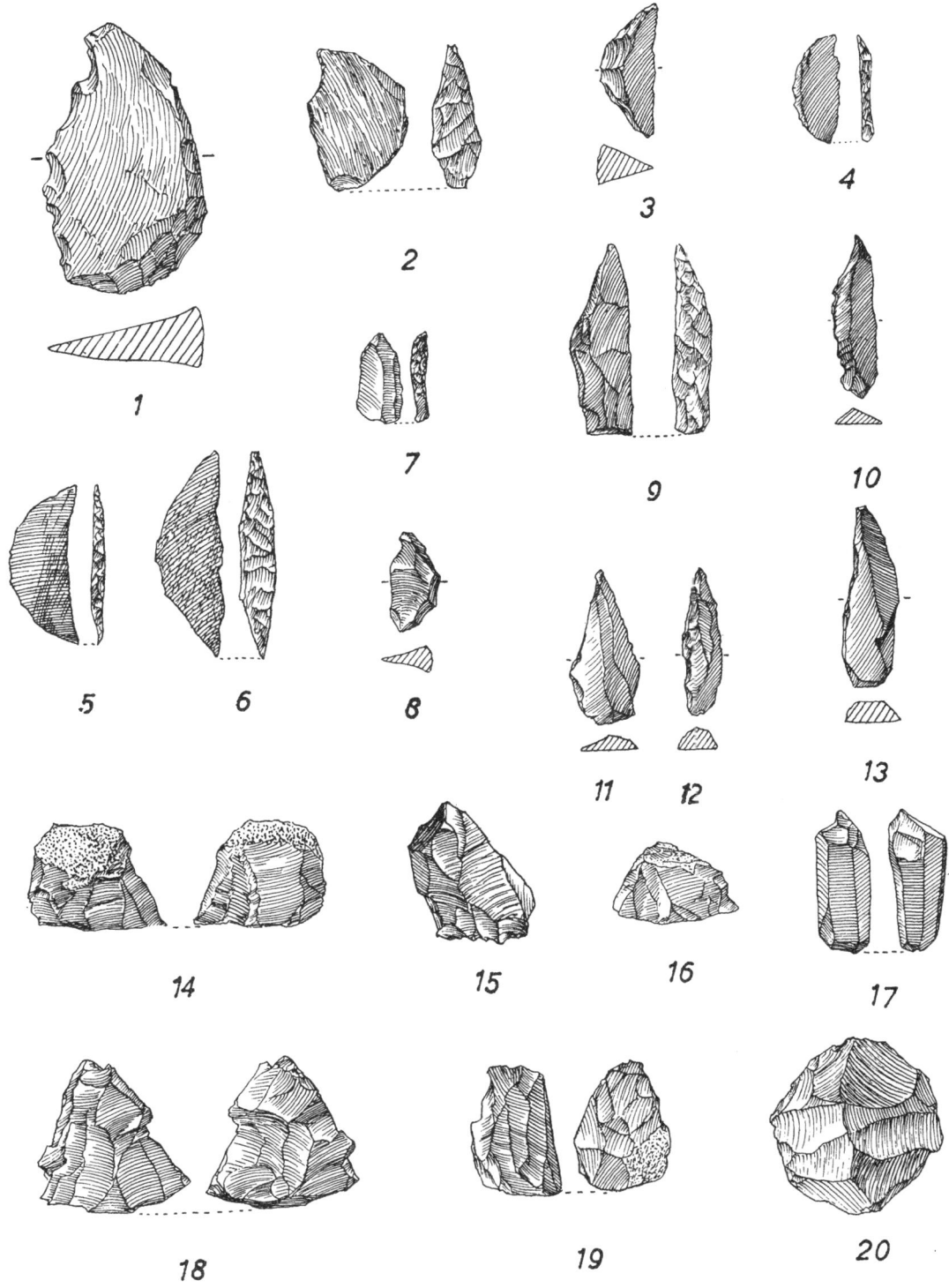

1 2 3 4 5 6 7 8 9 10 11 12 13 14 15 16 17 18 19 20

279

FIGURE 53

Neolithic

Specimens from the Wilton-Neolithic A industry at the rock-shelter at Nsongezi; white quartz and rock crystal.

Nos. 1–6: End- and thumb-nail scrapers of Wilton type.

No. 7: Schist polisher, perhaps used in burnishing pottery.

No. 8: Oyster shell bead.

No. 9: Ostrich egg-shell bead.

Nos. 10 and 11: Pebble cores.

$\frac{1}{1}$ scale.

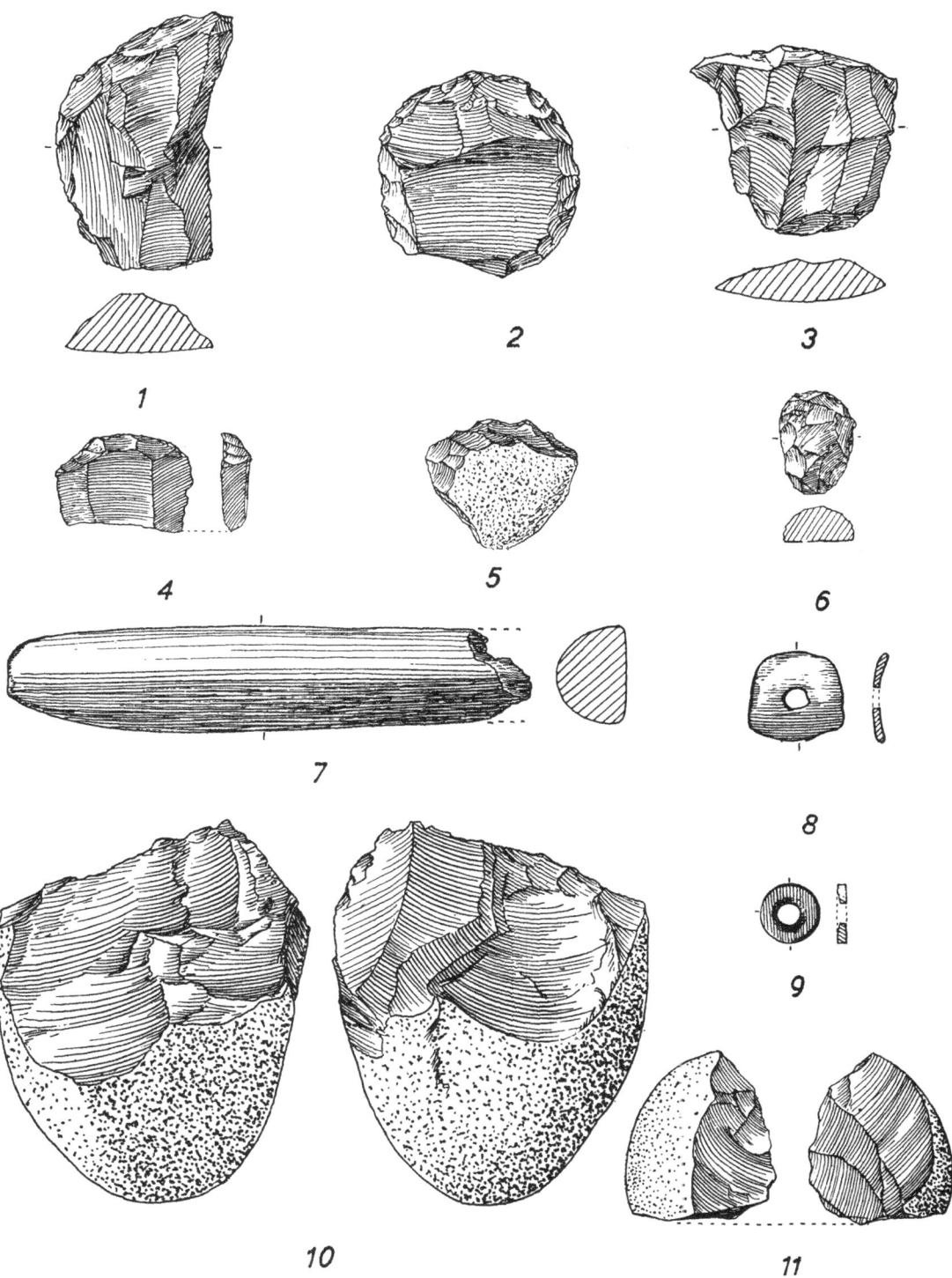

1

2

3

4

5

6

7

8

9

10

11

281

FIGURE 54

Neolithic

A bone anvil (?) from the Wilton-Neolithic A industry at the rock-shelter at Nsongezi.
$\frac{1}{1}$ scale.

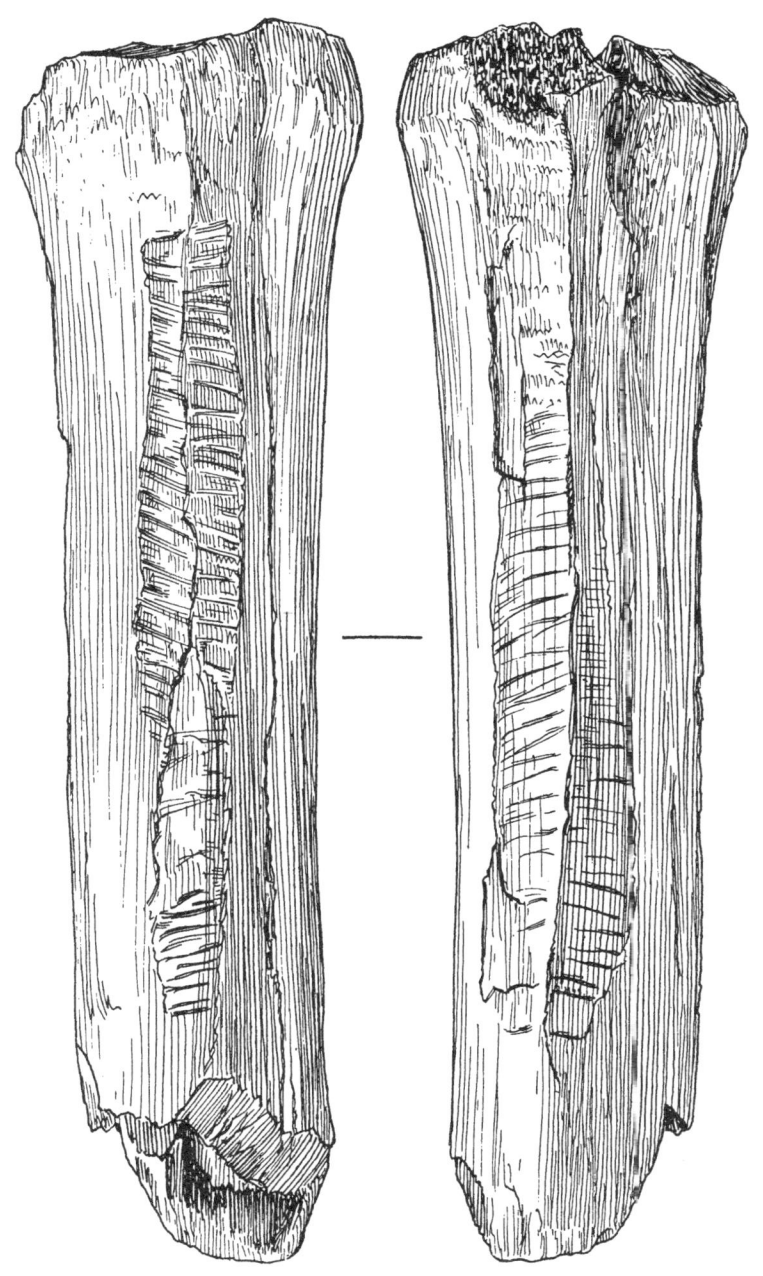

PLATE XXV

Wilton-Neolithic Culture

Rim fragments of decorated pottery from the rock-shelter at Nsongezi.

Figs. 1 and 2: from 2 ft. level.

Figs. 3, 4 and 5: from 2 to 3 ft. level.

Figs. 6, 7, 8 and 9: from 3 ft. level.

Fig. 10: from 3 to 4 ft. level.

All the sherds except Nos. 4, 8 and 10 are black burnished on both sides. No. 4 is burnished on the outside only, while No. 8 appears to be slightly burnished on the inside. No. 10 is of quite different fabric type from any of the other specimens: it is much softer, contains larger pieces of grit and bears no sign of surface polishing.

In all cases the decoration was carried out by light incisions or by drawing a narrow spatula round the rim of the vessel.

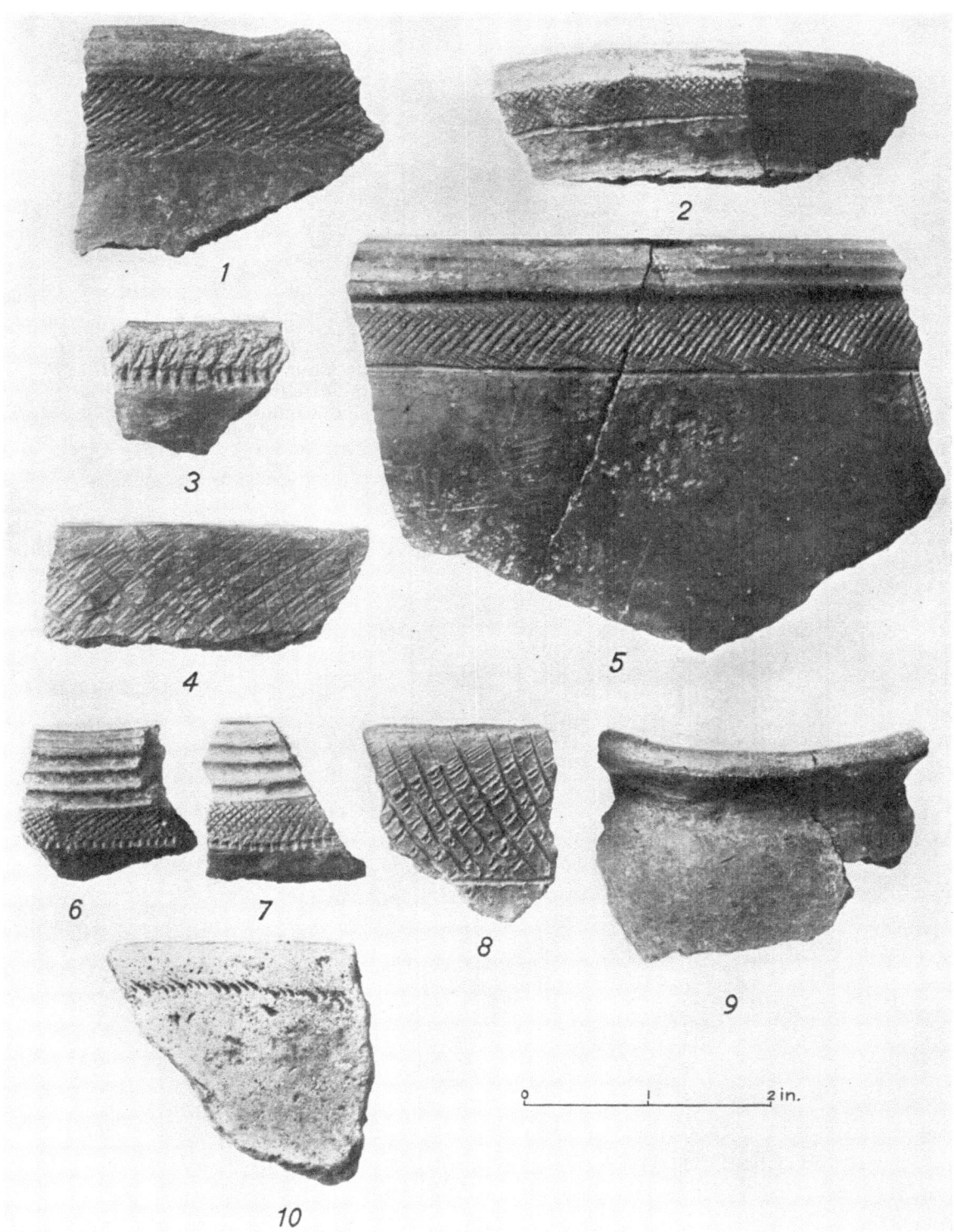

FIGURE 55

Neolithic

Specimens from the Wilton-Neolithic B industry in the Chui Cave, at Walasi Hill, near Mbale. Note the two gravers. The small figures indicate the levels at which the tools occurred. There was no observable difference in the industry throughout, so only tools from the top and bottom levels are figured.

$\frac{1}{1}$ scale.

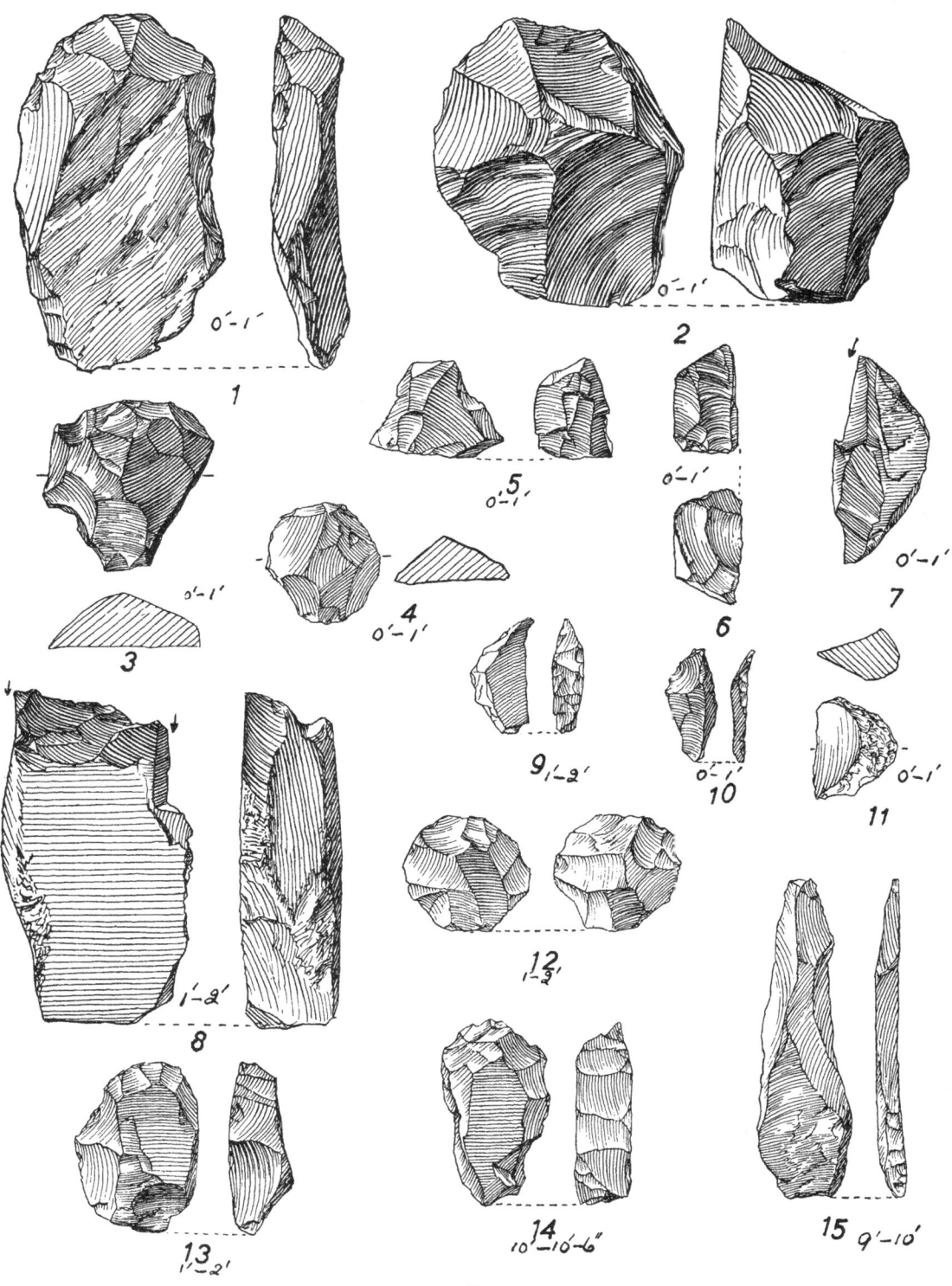

1

0'-1'

2

0'-1'

3

0'-1'

4
0'-1'

5
0'-1'

6

0'-1'

7

0'-1'

8
1'-2'

9 1-2'

10
0'-1'

11
0'-1'

12
1-2'

13
1-2'

14
10'-10-6"

15 9'-10'

287

FIGURE 56

Neolithic

Large stone objects from the Wilton-Neolithic B industry in the Chui Cave, at Walasi Hill, near Mbale. No. 1 is a piece of local lava which was probably used as a hand-guard when making fire by twirling a fire stick with a bow. No. 2 is a hammer stone and No. 3 is a pebble used as a grinder; we found no mortars or querns at this or any other home site in Uganda.

$\frac{1}{1}$ scale.

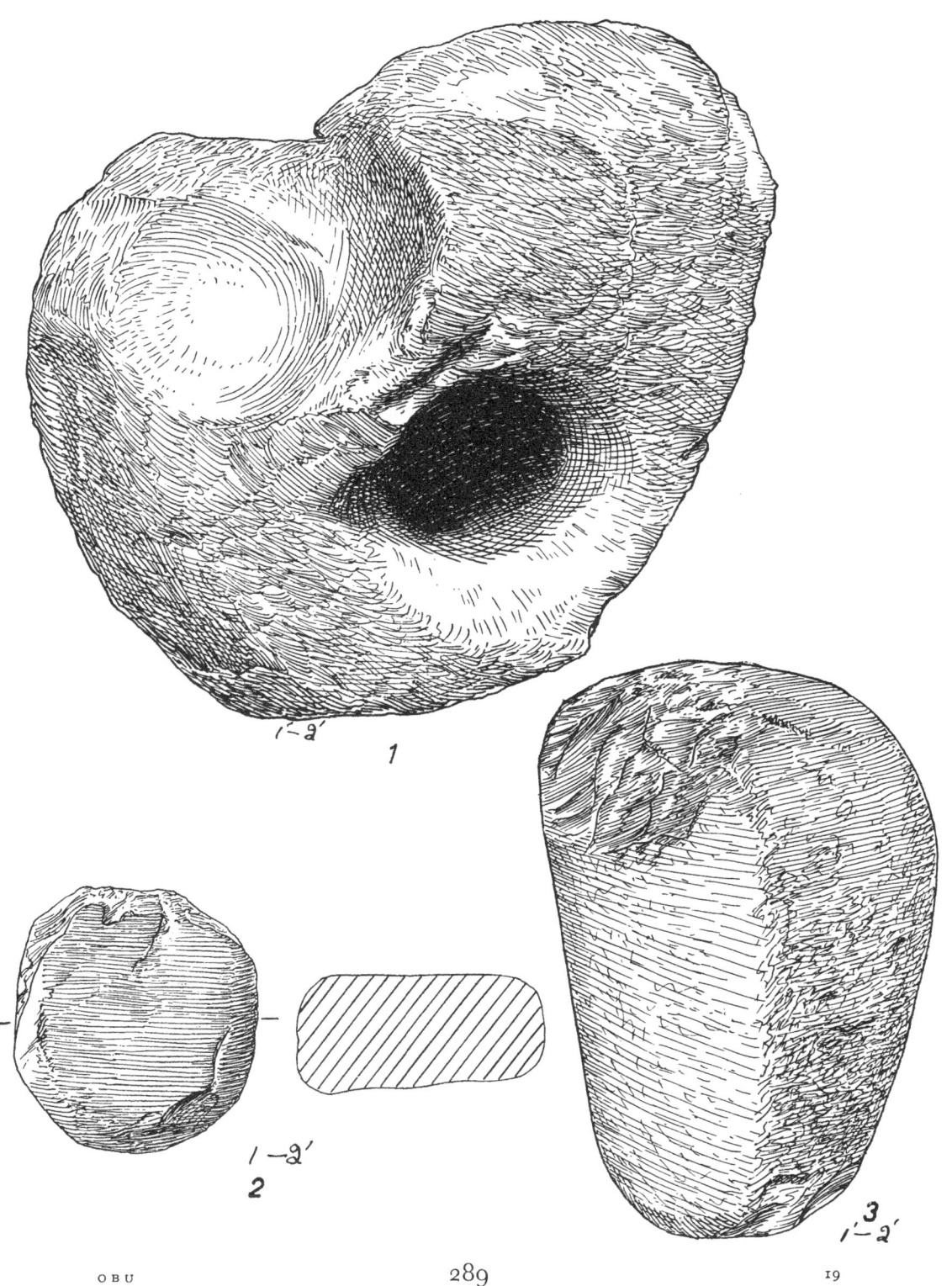

1

1 — 2'

2
1 — 2'

3
1 — 2'

PLATE XXVI

Neolithic Pottery

Fig. 1: top left. Neolithic pot in place in brickearth overlying a rubble at the Railway borrow pit, near Bugungu. The vessel can be seen resting on a flat piece of clay which also supported two other similar vessels behind the first.

Fig. 2: top right. Two aspects of one of the vessels from the site shown in fig. 1. Note the minute jab marks (pecking) on the side of the vessel in the upper picture.

Fig. 3: bottom left. Three similar pots found in wash earth just above the Muzizi river. The vessel in the centre can be seen resting on the remains of a similar clay platform as in the case of the Bugungu pots. The scale is in centimetres.

Fig. 4: bottom right. A microlithic tool—perhaps a borer—found with the pots shown on the left. Made in white quartz, sharp.

1

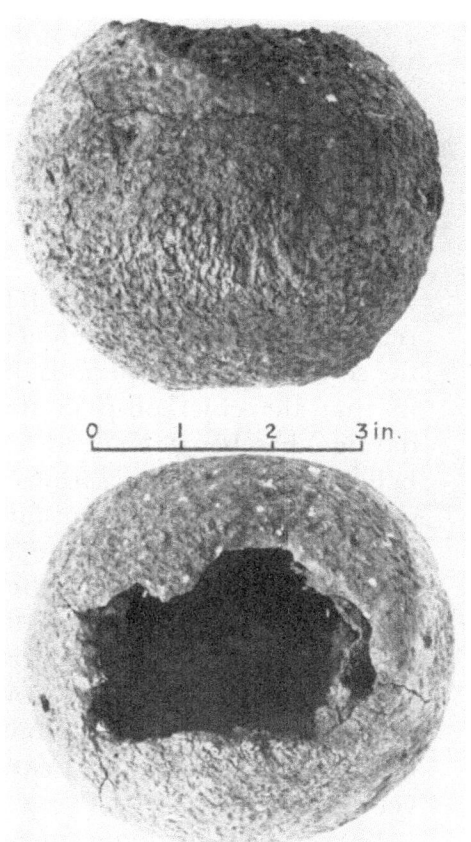

0   1   2   3 in.

2

3

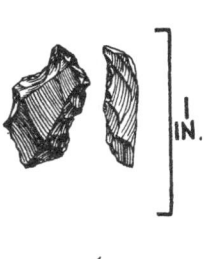

1 IN.

4

291

CHAPTER XIV

Correlations

FROM the point of view of the archaeologist, Solomon's conclusions regarding the lack of evidence of well-defined climatic events, such as the so-called pluvial periods, may seem to be in the nature of a tragedy, in that the chief prop in the East African prehistoric scaffolding—dating—has been removed or, at least, seriously weakened. I do not believe that this is a serious difficulty, however, simply because the significance of pluvials and interpluvials in East Africa has been greatly overrated. Apart from the doubts, felt by some of us, of the very existence of such periods, the pluvial scheme itself was not free from grave internal troubles regarding the number of the events, their degree of severity and, above all, from the archaeologist's point of view, their contemporaneity in such contiguous areas as Kenya and Uganda. From the first, it was obvious that considerable discrepancies existed between Wayland's and Leakey's schemes which rendered correlation very difficult.

Leakey was fortunate, in Kenya, in the almost superb simplicity of the Rift Valley lake-beds, laid, for the most part, in enclosed basins. In Uganda, the picture had to be reconstructed out of a jig-saw of gravel-patches, terrace-remnants, tilts, pot-holes, rubbles and other, frequently unrelated, phenomena and the wonder of it is that, after 17 years of constant work, checking and re-checking, correlations and hypotheses being many times changed, Wayland was able to come to even some measure of finality in his scheme published in 1934.[1]

In our opinion, many of the difficulties encountered in this investigation were due to the tendency of keeping the pluvial hypothesis constantly in mind and in attempting to relate all phenomena to it. It is surprising how many "normal" riverine or lacustrine deposits have been interpreted as of pluvial origin because of their position compared to beds forming to-day and because the prevalent belief that the pluvials matched the Pleistocene glaciations seemed to demand such an inter-

[1] See Table I, after p. 316.

pretation. Once the pluvial theory was questioned, however, and the relevant deposits were considered primarily as geological and not as meteorological phenomena, the whole problem became simplified and resolved itself into the query—could such and such deposits have formed in such and such positions during the normal course of geomorphology, or must we explain them as the results of climatic phenomena? In most cases, we found the former to be the simpler and more satisfactory explanation, as Solomon has shown. From this fact emerged the plain truth that we could not effect close climatic correlations even between such adjacent areas as Uganda and Kenya, let alone with Europe.

Turning to the archaeology, one would, at first sight, appear to be on common ground, at last. The cultures and industries so abundant in Kenya and Tanganyika surely could not fail to occur in Uganda, next door, and it should be easy to effect correlations. Yet, again, and in a manner most trying to the theorist, Uganda proved extraordinarily independent of her neighbour and, in spite of a multiplicity of lithic remains which are similar, in a generalised way, to those of Kenya, is obstinately incapable, at present, of close correlation, period for period. For instance, where, in Kenya, are there datable industries matching the Uganda Kafuan—the earliest of which is pre-Chellean and the latest of Upper Palaeolithic date? Again, while the Lower Uganda Oldowan matches the Tanganyika Oldowan both in date and typology, where, in Tanganyika-Kenya is there the slightest suggestion that this culture outlived the Chellean, which is there supposed to be its child, or that it ultimately merged with an Acheulean stage, as in Uganda? Further, I would say that the Uganda Chellean, with its huge flakes, resembles more the South African Chellean (Stellenbosch I) rather than any of the Oldoway stages, while, in the rest of the Tanganyika-Kenya Chellean and Acheulean there is surprisingly little to link with Uganda. Again, though the true hand-axe tradition of the Middle Uganda Acheulean is very similar to the Oldoway Bed IV Stages 3–4 (one cannot be more precise), the Late Oldowan element, so marked in the former, is entirely lacking in the latter, unless it has not been recognised. Nor do we appear to have, in Uganda, that Early Levalloisian mixed with Upper Acheulean, called Nanyukian, in Kenya. Then, with the exception of a single industry,[1] there is no Still Bay of Tanganyika-Kenya facies in Uganda but, instead, the Late Levalloisian appears to take on a new lease of life by contact with the Tumbian and to blossom, for a time, as the Walasi Still Bay Variation.

It is true that both Proto- and developed Tumbian stages occur in both Kenya and Uganda but, as far as I know, the Kenya Tumbian sites are all in the extreme west and obviously owe their existence to migrations from Uganda; the culture seems never to have penetrated farther east into the homeland of the Kenya Aurignacian, in the Rift Valley. Similarly, the Kenya Aurignacian, or Capsian, culture is entirely lacking in Uganda, and, though the two areas had a common culture in the Magosian, differences are observable in this, too.

There is, so far, no trace in Uganda of Late Kenya Aurignacian derivatives, such as the Elmenteitan and true Wilton,[2] but, on the other hand, Late Wilton combines with something else, perhaps Gumban, to produce the Uganda Wilton-Neolithic. There is no Njoroan, or polished axe culture so far discovered in Uganda, yet such tools are known in the Congo, the Sudan and in Kenya as far west as Lake Victoria. In Uganda, however, we have the curious, non-microlithic Kageran industry of just pre-Wilton-Neolithic age, which does not appear to be represented in Kenya.

The whole prehistoric complex in Uganda is different also from that in Kenya-Tanganyika in two other respects—environment and raw materials. I cannot agree with Leakey's opinion that the quality of the latter made no difference to Stone Age Man and that, for instance, industry *X*, when made in obsidian or some other close-grained lava, in Kenya, must be the same as when made of coarse quartzite or quartz, in Uganda. Nor can we believe that Stone Age Man was not sensitive to differences in climate and surroundings, at times preferring one region to another, even though his food supply were not seriously enough affected to cause his removal. It can hardly be doubted that many of the numerous Stone Age races inhabiting East Africa must have preferred the healthy, rolling uplands of Kenya to the low-lying swamps or thick forests of Uganda. Any slight increase in rainfall, such as occurred

[1] The Hoima industry.
[2] Without Magosian influence.

at various times,[1] must have accentuated the humid, tropical discomfort of the latter country, without making very much difference in the former. Then, as now, the boundary between the two countries was probably a natural one, involving the change from the airy uplands in the east to the steamy lowlands of the west—the same change as is felt to-day when one enters Uganda by road from almost any part of Kenya.

So it is that, in nearly every respect, the two areas had a differing human history, although, broadly speaking, the greater cultures were ubiquitous. Any attempt at close correlation between the irrespective culture stages fails at once, however, and much more work remains to be done before we can combine the two Stone Age sequences into a whole.

One important fact that emerged during the course of our work was that it is no longer possible or desirable for African archaeologists to turn their eyes northwards to Europe for guidance and enlightenment in their many problems. We can no longer regard Africa, at least south of the Mediterranean area of colonisation, as having been continuously influenced from the north and east, but must recognise that, apart from an occasional foreign impetus, African Stone Age Man achieved his own, essentially African culture, built out of African materials, in an African environment to which that culture was especially and purposely suited. In that realisation lies, I believe, a promise of tremendous significance for the future of African archaeology and for the solution of racial and kindred anthropological problems in that continent.

---

[1] What we call *pluvial episodes*.

APPENDIX A

THE M-HORIZON AND ITS CORRELATION WITH ACHEULEAN DEPOSITS IN KENYA AND AT OLDOWAY, TANGANYIKA

BY

T. P. O'BRIEN

RATHER than encumber the chapter on the Acheulean with a weight of detail bearing on the M-Horizon, I have written this appendix as a supplement to the relevant information in Chapter IX. Moreover, it is most necessary to examine in detail some claims and theories concerning the M-Horizon that have been put forward by Mr E. J. Wayland in *Bulletin* No. 2 of the Geological Survey, 1935, as these have an important bearing on the nature, age and duration of the horizon, and would, if correct, materially alter our conception of the period covering its formation.

The particular points of Mr Wayland's thesis which require comment are first summarised and then discussed individually.

(1) According to Mr Wayland the M-Horizon contains a mixture of tools which, pending his detailed study, he describes as *Chelleo-Acheulean, with a flake-tool development and derived Chellean, pre-Chellean and Oldowan forms* and that at Nsongezi *the latest of the tools represent one of the phases of the East African Acheulean.* Except for the derived tools *the artifacts are those of the people who followed the lake during its decline and retreated with the shore-line as it rose again with the rising lake.* Then follows more discussion and a table setting out the hypothetical changes that accompanied the falling and rising lake.

(2) Arguing from this initial assumption Mr Wayland goes on to assert that the M-Horizon *must be comprehensive with regard to steps of development in proportion as the time represented by the fall and rise of the lake is lengthy and in proportion as the site from which the tools are recovered is distant from the present lake.*

(3) *...investigations, so far as they have gone, reveal no great change in the contemporaneous* (with the horizon) *culture; nor does it seem that this condition of things is peculiar to the area...* (therefore) *it is permissible to infer that the time during which the M-Horizon was formed was relatively short; quite incompatible, at any rate, with the span required for an interpluvial period.*

(4) Mr Wayland then proceeds to correlate the M-Horizon with the reddened Bed III at Oldoway and with a certain deposit at the Kariandusi River

in Kenya, which latter contains *a similar assemblage of stone tools* and which is followed by *silts with intercalated ferruginous gravels particularly near the base*. The reddening of Bed III, M-Horizon and Kariandusi deposits is considered as proof of an oscillation to drier conditions but Mr Wayland contends that there is *nothing to show that desiccation supervened during the oscillation, although a phase characterised by long, dry seasons separated by pronounced wet seasons is by no means improbable.... There can be little doubt that a virtually rainless climate is not conducive to the production of red or ochreous deposits...if only because a certain amount of moisture is necessary to promote decomposition leading to the formation of iron oxides.*

(5) Mr Wayland points out that pebble beds occur at or near the top of Oldoway Bed III as well as in the M-Horizon and that some workers have assumed that these were a corollary of swollen streams. He justly adds that the exact reverse was actually the case, and he explains how, by the shrinkage of a swamp-lake, the effective zone of pebble deposition below incoming streamlets may move down with the diminishing water—standing water always acting as a velocity check on such affluents.

These points may now be dealt with in order, as follows:

(1) Mr Wayland has admittedly had greater facilities for the study of the M-Horizon than this expedition had, in that his geological investigations in the Kagera valley necessitated the digging of many shafts and workings. It is certain, therefore, that his collections from the horizon exceed ours in total number. But the horizon is largely a rubble formed on a land surface and, therefore, to some extent detrital, so that it is very likely that, if a very large number of artifacts were collected from it, a certain number of derived tools (from the rubbles lining the valley sides) might be found amongst them. We ourselves never found any such derived implements, however, and I emphatically reject the suggestion that the Oldowan forms to which Mr Wayland refers belong to such a group. In the Phase A, gravelly *facies* of the horizon they form the bulk of the core tools and, as I have stated in Chapter IX, are in exactly the same state of preservation as the Early-Middle Acheulean hand-axes from the same deposit: there can be no doubt of the contemporaneity of both classes of implement.

(2) In discussing the hypothetical cultural changes following the falling and rising lake Mr Wayland seems to have put up an Aunt Sally which he himself proceeds to demolish later. Its purpose seems to

have been an attempt to prove his contention that the M-Horizon was not of long duration.

(3) and (4) By denying the existence of several, if any, cultural changes, he argues that the M-Horizon must have been formed in a relatively short time. An understanding of this point of view is achieved when it is seen in (4) that the M-Horizon is no longer regarded as marking a very dry era. However, this is an assumption which is based on no particularly conclusive evidence. It must be conceded that moisture is necessary to effect the ferruginisation of such beds, but it must not be forgotten that not only is the M-Horizon in the 100 ft. terrace almost always underlain by a thick homogeneous bed of clay, but that it is followed by waterlaid sediments. These do not show much force of current action but seem merely to have covered up the stone bed as soon as normal rainfall returned to the region. But these deposits consist of sands and clays which are mainly of swamp type, and I suggest that the ferruginisation of the underlying M-Horizon most probably occurred at this later period owing to the downward seepage of iron-charged moisture derived from the swamps, until it reached the impervious clay stratum immediately below the stone bed. In some places this ferruginisation extends upwards into the lower part of the post-M-Horizon sands thus proving its post-M-Horizon date. But perhaps the best example of this mode of ferruginisation of underlying deposits from above is to be seen at the great section of sands, gravels and clays at the Mwirasandu Mine Hydro-electric Station at Kikagati, some eight or nine miles above Nsongezi. The section displays an unbroken succession from bottom to top. In it are several extremely hard, ferruginised sand horizons up to 2 ft. in thickness which are underlain either by clay or by fine-grained sands. They do not in any way represent breaks in the waterlaid succession and it is quite clear that their impregnation with iron precipitates took place either by the direct downward percolation of iron-charged waters from the overlying deposits or during periods when the water table sank by reason of partial desiccation within the valley. In any event, the fine-grained beds immediately below the ferruginised bands would seem to have acted as checks upon the further downward seepage of the iron solutions which thus had time to deposit their precipitates.

Appendix A

Apart from the actual ferruginisation of the M-Horizon, there is a strong possibility, in my opinion, that its induration does not take place until the bed is exposed in ravines or, at any rate, brought near the surface by erosion of the succeeding beds. This was suggested on it being noticed that when pits were dug through the thick upper beds to the horizon, it was never indurated or even strongly compacted, although heavily ferruginised. I suggest that the free access of air or surface water containing oxygen may be necessary before actual induration of such deposits can take place. Many people with experience of tropical laterite will know that when it is freshly exposed it frequently hardens considerably, presumably by the dehydration of the iron contents.

Mr Wayland's explanation, that the ferruginisation of the M-Horizon took place at the surface at the time of its formation as a rubble, is extremely unlikely for it necessitates a semi-waterlogged condition operating for a long time and that, in view of the very nature of the deposit—a land-surface accumulation that was fairly well drained by its position on the sloping sides and floor of the valley—seems an impossible event.

It is not to be questioned that the formation of "ironstone" most frequently, if not always, takes place in marshes or shallow lakes where, by the decay of organic matter, iron is removed from the deposits and thrown down either as a carbonate or as a hydrous ferric oxide.

In this connection, one is tempted to ask whether the reddening of Bed III, at Oldoway, was contemporary with, or later than, its deposition. Apart from its striking colour, nobody doubts that it is anything more than the continuation of Bed II, for it carries the same minerals and general volcanic debris: its colour alone (apart from the contemporary human culture) distinguishes it from Bed II. It is, however, followed by a bed which yields quartz for the first time in the succession and a large percentage of land fauna, mainly antelopes. I am aware that these two factors, when treated as indices of climate, seem to be contradictory, as Solomon has remarked (p. 41), but I do not think that this is necessarily true until it is proved that the quartz grains really represent increased stream velocity by bringing in minerals derived from the ancient crystallines of the surrounding country. I think it is probable that Mr Wayland's own explanation of the bands of pebbles in the upper part of Bed III, due to downward shifting current velocity checks and the growth of minor deltaic fans, may equally well account for the incoming of quartz grains in Bed IV. If we believe that the Bed III pebble zones represent drier periods (perhaps only on a seasonal scale) and, further, that the general climatic trend was towards still drier conditions, we may regard the sandy constituents of Bed IV as having been deposited in the same way by the outward and downward growth (towards the lower parts of the basin) of subaqueous, topset, delta accumulations. Apart from a purely water-borne origin of the Bed IV quartz grains, however, we may ask whether it is not possible for wind erosion over an increasingly arid topography to have led to their deposition in the dying Oldoway marshes. This brings one to the question of the reddening of the Bed III sediments. Mr Wayland assumes that it was contemporary with the actual deposition of the bed. He quotes Twenhofel in showing that the reddening of such beds (by the oxidation of iron compounds) is most usually accomplished during alternating periods of dryness and humidity, heat being a considerable factor in accelerating the process of decomposition. I have already dealt with this point of view as applied to the M-Horizon and shown that, in my opinion, a more likely explanation of the ferruginisation there is that it was a secondary effect, which took place *after* the formation of the horizon. In the same way, I suggest that the reddening of Bed III, far from indicating the condition of the climate during the bed's formation, may actually have taken place during the formation of Bed IV, when, by the time of the Middle Acheulean, the climate of East Africa was much—if not very much—drier than before.

The evidence of the M-Horizon insists upon an important dry period at this time and it may be considered reasonable to suppose that some indication of its effects should be found in the Oldoway Series. As I will show shortly, there is no possibility of believing that the Bed III and M-Horizon stages of Acheulean are the same; the latter is Middle and not, as Mr Wayland thinks, Early Acheulean. I suggest, therefore, that the dry period responsible for the M-Horizon is represented in Oldoway Bed IV both by the increased percentage of antelope forms and by the incidence of detrital quartz, and, further, that the reddening of Bed

III was not a climatic factor but took place, as it did in the M-Horizon, by the downward seepage of iron solutions in water derived from the overlying sediments, i.e. Bed IV.

These considerations must tend to reduce the value of ferruginisation in deposits like Oldoway Bed III and the M-Horizon as an indication of the meteorological conditions at the time of their formation, provided that they were followed (as these are) by "wet" deposits, and we are forced to try to estimate the climatic factors by other means. There is no cultural break either between Beds III and IV or in the latter, such as one might expect if the prevailing climate had affected Man in any way. The probable explanation for this is that the Oldoway swamps lay in an enclosed basin, receiving the whole of the local drainage, and so they did not dry up to such an extent (or at any rate, so rapidly) as did the open Kagera valley, some eighty miles from the main Victoria basin.

As I have already shown in Chapter IX, the very concentration of Middle Acheulean tools in the M-Horizon suggests that the people were forced to congregate where they could, perhaps, obtain water at rain pools or by digging. It is most significant that, in spite of the almost unbelievable richness of the M-Horizon in the 100 ft. terrace and the comparative abundance of both earlier and later industries in the hillside rubbles, we could not find more than a single one or two typical M-Horizon specimens outside the terrace, even at places a short distance up the valley sides. Moreover, as Mr Wayland has made a special point of the comparatively slight changes to be seen in the M-Horizon Acheulean, and argues that this must be proof of its short duration as a land surface, it may be noted that this fact was also apparent to us but that it and the further fact of the enormous concentration of these tools in the valley bottom, only suggested the severity of the drought without giving any clue to its duration. This clue, however, is supplied by the tools in the succeeding waterlaid beds. These post-M-Horizon deposits are rich in several industries, utterly different from the M-Horizon assemblage, and this fact demonstrates a most important culture break in the succession. If, as Mr Wayland thinks, that the M-Horizon only lasted a comparatively short time, and there was no intervening break, we should expect a continuation of the Acheulean in the succeeding beds. In actual fact, there is no trace of hand-axe tools until the Proto-Tumbian industry is met with in the N-Horizon, but the intervening 17 ft. ± of sand contain several horizons of typical East African Levalloisian—a culture never found in deposits earlier than the Gamblian.

All these considerations go to show that the period represented by the M-Horizon was both arid enough eventually to have driven Acheulean man to more favoured areas, such as the great lakes, and long enough for that culture to have died right out before the next series of deposits was laid.

I agree with Mr Wayland that the period covered by the Middle Acheulean industry in the M-Horizon was probably not very long, but I do not agree that sedimentation began again soon after the stone bed formed. In other words, the M-Horizon not only represents the period of Middle Acheulean occupation of the valley, but also some unknown lapse of time afterwards, right up to the oncoming of "Gamblian" conditions and an influx of fresh culture.

Turning now to the correlation which Mr Wayland effects between Oldoway Bed III, the M-Horizon and the Kariandusi River Acheulean deposits, we must examine the evidence for or against its validity, particularly in view of Professor van Riet Lowe's agreement with Mr Wayland that the Bed III and M-Horizon industries are identical.[1]

I have no personal knowledge of Bed III, and the tools that I have seen from it are only small museum collections, of which the best is probably that at the Coryndon Museum, Nairobi. I have, however, first hand knowledge of both the M-Horizon and Kariandusi deposits, which latter Dr Leakey kindly gave me leave to study. Opportunities of doing so were presented both on our way to Uganda in 1934 and on our return in 1936.

It will be remembered that in *The Stone Age Cultures of Kenya Colony*, Dr Leakey spoke of both rolled Chellean and unrolled Acheulean tools from the Kariandusi River. Subsequently, in *Stone Age Africa*, he says that this view was mistaken and that the tools that he had previously regarded as Chellean are, in fact, merely "*unfinished specimens from the*

---

[1] Appendix A, *Geology and Archaeology of the Vaal River Basin*, Memoir No. 35, Dept. of Mines, South Africa, 1937.

factory site...of the time of the fourth stage of the Acheulean". He does not, however, explain the instantly observable fact that the vast majority of the "unfinished" tools are heavily water-rolled and that the "finished" ones are fresh or very much less rolled! Considering that the site in question is on the bank of the river and that the main implementiferous deposit is a gravel, whose deposition, in Solomon's opinion (Appendix A, *Stone Age Cultures of Kenya Colony*), was connected with the initiation of the river at this time, it is difficult to understand how the place could possibly be regarded as a factory site; the rolled tools interbedded with the gravels can only have come from some distance away. On the other hand, we observed that the majority of fresh or only slightly rolled artifacts occur in a rubbly layer on the surface of the main gravel. In some cases, however, single, fresh, well-made specimens occur in the latter and I suggest that these have slipped down root holes or other crevices and that their apparent dating of the gravel is utterly deceptive. I think that there can be no doubt that two distinct stages of the Acheulean are present, the rolled, cruder implements belonging to the earlier, and the fresher, better-made ones to the later.

After examining the Oldoway series at the Coryndon Museum, I entirely agree with Dr Leakey's dating of the fresher Kariandusi specimens to the fourth stage of East African Acheulean, but I feel that the earlier, rolled stage more closely compares with Stage 3 at Oldoway. From the point of view of equating the deposits at the Kariandusi River with Bed III this dating is of importance, for the latter contains only Early Acheulean, or Stage I, and Mr Wayland's tentative correlation between these deposits, when based on their contained industries, is invalidated. His reference to the post-Acheulean deposits at the Kariandusi River, containing ferruginous gravels, is calculated to support the correlation between such "red" deposits, but it must be pointed out that these particular ferruginised gravels occur as intercalations in *Gamblian* silts, producing Upper Palaeolithic remains, which were only deposited after some considerable lapse of time and after much erosion of the Kamasian deposits had taken place. Thus both archaeology and geology agree in upsetting the proposed correlation.

I come now to the equation Oldoway Bed III = M-Horizon. At the beginning of Chapter IX I showed how we had at first followed

Mr Wayland's lead in regarding the M-Horizon as of Early Acheulean date, but that study of the Coryndon Oldoway collections supported our tentative belief that it was of later age, i.e. Middle Acheulean of about the standard of Oldoway Stages 3–4. Still further study in England of the relevant industries has not caused any change in this view. I can only repeat that I consider the Bed III material to be altogether cruder than even the Early-Middle Acheulean (Phase A) in the M-Horizon, while both M-Horizon stages closely resemble the two stages at the Kariandusi River. The only significant difference in the typology of the industries at the two sites is the great percentage of inherited Oldowan core chopper forms in the Uganda group and the somewhat smaller average size of the Kenya tools when a large series is examined. On the other hand, there is a great difference between the tools of Bed III and the M-Horizon, especially in the great abundance and variety of cleaver forms in the latter, where, indeed, they form nearly half the total number of Acheulean types, whereas in Bed III they are rare. Leakey, in fact, regards their first appearance there as one of the proofs that the Chellean had at last given place to the Acheulean. In view of the apparent supreme perfection of the Chelleo-Acheulean evolution in the Oldoway deposits, it can hardly be credited that such specialisation of the cleaver form as occurs in the M-Horizon should be as early as its *earliest* appearance in Bed III. Professor van Riet Lowe comments, in his recent work,[1] on the Bed III cleavers and remarks that he does not consider this implement to be a significant factor in determining the earliest appearance of the Acheulean in East Africa, apparently because in South Africa it is found as early as the Chellean (Stellenbosch I). This opinion is understandable as long as one considers the whole African Chelleo-Acheulean culture to be a single, indivisible complex and does not take into account the modifications imposed on local industries by environment, raw material and so on. I believe that small differences can exist in contemporaneous industries even in such closely connected regions as Kenya and Uganda, where, in fact, there are few stages that do not show local variations due to one or other of the causes that I have suggested. Furthermore, small variations between contiguous areas may easily become big differences between regions as

[1] *Op. cit.* p. 125.

far apart as East and South Africa. For this reason there need not be any cause for surprise that the cleaver does not appear in East Africa until the Acheulean. But when, as I have remarked, one considers the unbroken evolution of the Oldoway Series, in which this tool type first appears in the Bed III Early Acheulean, one cannot but believe that a considerable time elapsed between this and the M-Horizon stage, when the cleavers are in equal numbers with the hand-axes and of several different types. The respective areas are far too close for there to be such important typological differences in industries of the same age.

Sufficient has been said to indicate that the proposed correlation between the "red" deposits of Oldoway Bed III, Kariandusi River and the M-Horizon, and their contained industries rests on slender foundations which, if critically examined, contain considerable factors of discrepancy.

(5) The last point of Mr Wayland's thesis deals with the incidence of pebble horizons in Bed III and the M-Horizon and their mode of formation. As regards Bed III, I have no personal knowledge of these pebble bands, but I am prepared to accept Mr Wayland's explanation that they probably represent small deltaic fans extending out over earlier silts in consequence of a diminishing water area and resultant shifting velocity checks opposite the mouths of affluent streams. I do not even challenge this explanation when applied to the earlier, gravelly *facies* of the M-Horizon. My own explanation of the latter, in Chapter IX, was that the gravel probably formed on the bottom of the flooded valley at the time when the valley-lake stood at the level of the beach at Mile 14 on the Nsongezi-Mbarara Road. This correlation, of course, depended on the similarity of the rolled Early-Middle Acheulean artifacts in the beach and valley gravels. I assumed that while the lake stood at this level a certain amount of lateral erosion round the valley sides occurred and that some of the tools of people living at or just above beach level were washed down the rather steeply sloping valley sides, to accumulate on its floor. Then, when the lake retreated down the valley, this gravel bed became a land surface used by the makers of the Phase B Middle Acheulean tools.

On the whole, I think Mr Wayland's explanation the better, though I disagree with its implication that the gravel could form, in places, at all times during the dry oscillation. The reason is that the tools from the gravel are always of an earlier type, and that the better-made Phase B tools are always in a true rubble overlying the gravel and are fresh. Assuming, therefore, that the gravels were formed as Mr Wayland suggests, this occurred during the time when the lake was retreating down the valley as a result of oncoming aridity, but before the valley bottom became a land surface. For all I know to the contrary, this same explanation may account for the circumstances obtaining at the Kari-andusi site, for there also, the earlier, implementiferous gravel overlies true lake silts (conformably) and is in turn overlain by a rubble containing a later industry.

CONCLUSIONS

The relevant facts and conclusions advanced above are here summarised briefly.

(1) The M-Horizon is typically composed of two parts containing Middle Acheulean industries of slightly different ages: Phase A, an earlier, gravelly *facies* containing an Early-Middle Acheulean industry, roughly contemporaneous with the industry associated with the lake beach near Nsongezi, and Phase B, a younger, rubble *facies* containing a developed Middle Acheulean industry. In Phase A the greater proportion of total artifacts consists of Upper Oldowan core-choppers in intimate association with the Early-Middle Acheulean industry. In Phase B these forms have become completely merged, the core-choppers being an integral part of the Middle Acheulean.

(2) The formation of both parts of the M-Horizon was the direct result of a climatic change from what we regard as a "normal" climate of moderate rainfall to a very dry one, sufficient to cause a considerable drop in the level of Lake Victoria. In the earlier part of this period, when there was still enough precipitation to effect stream action the gravelly *facies* of the horizon came into existence. Somewhat later, after the lake waters had retreated below Nsongezi and stream action ceased, the rubbly Phase B was formed on the newly exposed gravels or mud-flats. From this point onwards there is no sign of further sedimentation within the limits of the horizon, though there probably was a certain amount of blown sand accumulation during the height of the dry epoch. Middle Acheulean Man ultimately left the area altogether before this latter period supervened. From then onwards, until the return of "normal" climate and the renewal of sedimentation, the region was unoccupied. Thus the M-Horizon must be considered to represent not only the Phase A and B depositional *facies*, but also, by inference, some unknown lapse of time between the Middle Acheulean and the overlying Gamblian deposits containing Levalloisian and other industries. In Kenya the comparable period was covered by Upper Acheulean industries (found in rubbles, as on the Kinangop Plateau) and at Oldoway by the same culture stages in the upper part of Bed IV.

Appendix A

(3) In the Kagera Valley the new sediments swept over and covered the M-Horizon stone bed, probably incorporating the blown sand deposits of the dry epoch. These post-M-Horizon current-bedded sands contain abundant Levalloisian artifacts and are dated to the early part of the Gamblian period, being followed in turn by later Gamblian sediments containing Proto-Tumbian, Tumbian and younger Levalloisian assemblages.

(4) The ferruginisation of the M-Horizon stone bed (Phases A and B collectively) probably took place at some period during the Gamblian by downward seepage of iron solutions from the overlying deposits, assisted by the decay of organic matter (marsh vegetation, etc.). These solutions could not easily penetrate the heavy clay stratum below the M-Horizon which thus became saturated with their precipitates. It is suggested that the induration of the ferruginised stone bed may have been a still later result depending on the access of oxygen after some erosion of the overlying sediments had taken place, after the river fell to the 30 ft. terrace level, or later.

(5) Neither the archaeological nor geological data support the correlation of Oldoway Bed III with the Kariandusi Acheulean beds and the M-Horizon. On the other hand, we consider the two latter to be approximately contemporary deposits, both of Middle Acheulean age and, therefore, equivalent in age to part of Oldoway Bed IV.

(6) It is not considered possible yet to correlate precisely *individual* culture stages in places so far apart as South and East Africa. Individual stages such as the Oldoway Bed III, the two Kariandusi and two M-Horizon stages must be considered on their own merits as belonging to an East African branch of the African Acheulean. Their likenesses or differences must be taken into account in a limited, regional sense and not as necessarily applicable to the whole African complex. Proof of this is seen in the considerable diversity in the African Upper Acheulean stages as between South Africa, where the Victoria West and Fauresmith cultural trends form distinct local groups, and East Africa, where the cultural evolution was somewhat different and standardised to almost European pattern.

APPENDIX B

THE MAMMALIAN FOSSILS

BY

A. Tindell Hopwood, D.Sc., F.L.S.,

Department of Geology, British Museum (Natural History)

Identifiable bones and teeth of mammals were found as fossils at three localities, namely, at Kaiso, along the banks of the Kazinga Channel, and at the Kikagati hydro-electric site. All these deposits are regarded as being of Lower Pleistocene age (cf. Hopwood, 1935, 1937). Unfortunately, there are no complete skulls or mandibles in the collection, and the material is not sufficient to enable one thoroughly to revise the account of the Kaiso mammals published over twelve years ago (Hopwood, 1926). That account was written before the great collections from Africa in London and Paris had been made, and before the study of those in Berlin and Munich had been completed. Now it is desirable to obtain additional material in order that statements then made, and which to-day appear to be doubtful, may be confirmed or refuted.

The paragraphs which follow are in the nature of a preliminary report; the material will be fully described in a catalogue of the Lower and Middle Pleistocene mammals of East Africa which is in preparation. The report is arranged in the same manner as an earlier one (Hopwood, 1931), that is, each locality is treated separately, the specimens being arranged in zoological order. I have also added specimens from higher horizons, and others which have reached the Museum from other sources, notably from E. J. Wayland, Esq., Director of the Geological Survey of Uganda. After each specimen is quoted the number it bears in the Departmental Registers. Names of persons followed by a year refer to the List of Literature at the end.

(i) LOCALITIES OF KAISO AGE

Kaiso

The specimens from Kaiso constitute by far the greater part of the collection, but most of them belong to *Hippopotamus kaisensis* Hopw. The remainder of the specimens have been identified as *Phacochoerus* sp., *Hippopotamus imaguncula* Hopw., *Stegodon kaisensis* sp.nov., and *Simopithecus* sp.

Phacochoerus sp. A wart-hog is represented by an imperfect third molar of an old animal (regd. M 15161). The tooth is considerably worn, and much of the hinder part is missing; it is insufficient for comparison with the teeth from Olduvai (Dietrich, 1937).

Hippopotamus kaisensis Hopw. (? = *H. gorgops* Dietrich). Represented by parts of at least two skeletons and a few teeth. No skull of this species has yet been obtained from Kaiso, but a skull (regd. M 15162) collected by Dr Leakey from beds of Kaiso age at Rawi, Kenya Colony, is indistinguishable from skulls of *H. gorgops* from Olduvai.

Details of all the material from the various localities will be published elsewhere, but the variation in the size of the teeth is illustrated by the measurements of some third lower molars taken at random. Unfortunately, there are no third lower molars from Rawi:

Regd. No.	Locality	Age	Length mm.	Breadth mm.	Height mm.	Index
M 12648	Kaiso	Kaiso	77	40·5	55·5	53
M 15163	Kaiso	Kaiso	75	41·5	55	55
M 15164	Kagua	Kaiso	89	48	—	54
M 15165	Kagua	Kaiso	88	50	—	57
M 15166	Olduvai	Bed II	77	46	60	60
M 15167	Olduvai	Bed II	71	41	55	58
M 15168	Olduvai	Bed IV	69	36	55	52

Hippopotamus imaguncula Hopw. A few teeth (regd. M 15169) were found in the same place as the skeletons of *H. kaisensis*.

Stegodon kaisensis sp.nov. A small species of *Stegodon* with eight or nine ridges in the third lower molar, each ridge made up of from seven to ten conules; enamel thick, rugose, triate, and grooved externally. Ridges close together, four to five in 10 cm., and low, 45–50 mm.

This species is founded upon an incomplete third lower molar (Holotype, regd. M 15170 B.M.G.D.) and a portion of another tooth (Paratype, regd. M 15171 B.M.G.D.). The holotype is broken; all that part of the tooth supported by the anterior root, together with the root itself, has been lost. Of the remainder, the last three ridges and talonid are half-worn; the anterior ridges are worn flat. Nevertheless it is possible to make out the former existence of six ridges on the specimen as it now is, and since the anterior root supported two, if not three ridges, the total number was probably eight or nine. The estimated length of the holotype when complete was about 250 mm. (present length 200 mm.); breadth at the eighth ridge 75 mm.; estimated height 35–40 mm. The corresponding figures for a tooth referred to *S. insignis* (regd. M 3038) are, length 270 mm., breadth 80 mm., height 52 mm., 3–3½ ridges in 10 cm.

Simopithecus sp. A piece of a maxilla with the right second and third molars (regd. M 15172) is referred to this genus. The teeth are much worn, and the enamel is chipped in places, nevertheless, the specimen is interesting as the first Primate to be recorded from Kaiso.

Kazinga Channel

The fossils collected from deposits of Kaiso age along the Kazinga Channel were obtained from two horizons. Those from the lower bed are indistinguishable from specimens found at Kaiso itself, and, to judge from the matrix, the lithology of this bed agrees with that of the rocks at Kaiso. The upper bed[1] is composed of soft, light yellowish sands with very little cement, but the contained fossils seem to be just as highly mineralised as those from Kaiso. Apart from a skull of *Hippopotamus imaguncula* Hopw. already in the Museum, none of the specimens from the lower bed has been identified with certainty.

Phacochoerus sp. Portions of four large tusks (regd. M 15173) seem to belong to a wart-hog.

Hippopotamus imaguncula Hopw. A skull, lacking the brain-case, the snout, and the zygomatic arches (regd. M 14801) has the cheek teeth from P4 to M3 on either side finely preserved; it was presented to the British Museum by the Director of the Geological Survey of Uganda. The summits of the premolars are just touched by wear; the first molar is greatly worn; the second molar is nearly half-worn; the third molar

---

[1] About 6 ft. above the lower bone horizon. There are usually several such horizons in the upper part of the Kaiso Series. T. P. O'B.

is not yet fully erupted, and is unworn. The length of the three molars is
100 mm.; their individual measurements are:

	Length mm.	Breadth mm.	Index
M1	31·5	31·5	100
M2	36·7	37·8	103
M3	35	36	102·9

The width of the palate at P4 is 46 mm., and at M2 44 mm.

Taurotragus sp. A mandibular ramus of a large subadult antelope
(regd. M15174) probably represents a species of eland.

Stegodon kaisensis Hopw. A single ridge of an unerupted tooth (regd.
M15175).

Kikagati Hydro-electric Station

Archidiskodon griqua Haughton. Mr O'Brien brought back four incomplete molars of an extinct elephant from this site (regd. M15210–
M15213); another (regd. M15209) had been received previously from
Mr Wayland. The specimens are important because, not only do they
prove that the deposits in which they were found are of the same age as
those at Kaiso, but also because they show that the specimens previously recorded from Kaiso under the names *Elephas zulu* Scott and
Elephas aff. *meridionalis* belong to one and the same species, which is not
E. zulu.

The most important piece of the new material is an incomplete third
lower molar (regd. M15211) with eleven plates remaining. Its greatest
length is 240 mm.; the greatest height is 110 mm. at the centre plate;
and the greatest width, measured at the grinding surface, is 68 mm.
Each plate has five digitations, except the sixth which has only four, and
the last but one, which has three. The plates increase in width from the
summit to within about 20 mm. of the base, from which point they tend
to become narrower again; the increase in thickness from front to back
is about 50%, from 10 to 15 mm. the clefts separating the two outer
digitations from those in the centre of the plates are deep and roughly
parallel, so that the increase in width of the plates is brought about by
an increase in width of the outer digitations. When first entering into
wear, the plates give the impression that the enamel figures will

develop from two lateral annuli and a median lamella (cf. Hopwood, 1926, p. 35, pl. iii, fig. 1), but in consequence of the arrangement of the digitations the enamel figures of the lower teeth actually develop from three sub-equal rings. The various stages are well shown in another imperfect lower tooth (regd. M15213).

Of the rings which compose the enamel figures, the two outer ones are narrow, parallel-sided bands of smooth enamel, whereas that in the centre is irregular, with one (regd. M15211) or two (regd. M15213) simple festoons on the anterior and posterior surfaces. When the rings coalesce, the resultant enamel figure is a narrow, parallel-sided band, sometimes a little bent forward, with one or two bold festoons of enamel in the centre. Apart from the festooning, the enamel is smooth and fairly thick—2·5 mm. on the anterior surface and 3·5 mm. on the posterior surface of the plate.

The upper dentition is represented by two pieces. One (regd. M15209) consists of the first four plates and anterior talon of a tooth which had only just begun to come into use; the other (regd. M15210) consists of the last five plates and talon of a worn tooth. The greatest length of the first specimen is 110 mm.; the greatest height, at the last plate, 108 mm.; the maximum width, at the base, 86 mm. Each plate has five or six digitations, and gradually increases in width from the summit to the base. The increase in thickness is the same as in the lower teeth, namely about 50%, or from 9 to 10 mm. at the summit to more than 13 mm. at the base. The clefts between the digitations are deep, but the two outer ones are deeper than the others and converge as they pass downwards, hence the enamel figures which might show at first as a medial lamella flanked by lateral annuli would gradually become three subequal rings, and finally a central annulus with a lateral lamella on either side. So far as one may judge, the entire enamel figure would not be visible until some 70 mm. of the fourth plate of M15209 had been worn away; it would have the form of a parallel-sided plate with festoons of enamel in the centre.

The second example of an upper tooth (regd. M15210) is damaged. Its maximum width is 71 mm., and there are four plates in 10 cm. The last plate but one, exclusive of the talon, is 61 mm. high; the digitations are worn off, and the enamel shows three sub-equal rings, that in the centre being slightly wider than the others. The plate in front of it is only 50 mm. high and the enamel rings are unequal; that in the centre is 17 mm. wide, whereas the others are 26·5 and 22 mm. wide respectively. In front of this plate is another, much broken, which was not more than 40 mm. high, and this is the first to show the rings uniting as a complete figure, which appears to have been a parallel-sided band with a median expansion. As a general rule, such deep clefts and corresponding delay in the formation of the complete enamel figures are only found in abnormal teeth, but in this species they appear to be but a slight exaggeration of the normal condition.

Compared with the specimens previously known from Kaiso, the specimens from Kikagati are intermediate between the fragments described as *Elephas*, cf. *meridionalis* and the lightly worn molars referred to *E. zulu*. One of the latter (regd. M12639) was cut parallel to the grinding surface so as to show the complete enamel figures, which are curved forwards, are somewhat irregular with subparallel sides, and have median sinuses resembling those in the teeth from Kikagati and in the fragment compared with *E. meridionalis*. Moreover, the specimens form a series ranging from M15211 with parallel-sided plates and more complex sinuses, through M15213 and M12641, to M12639 with more irregular plates and simpler sinuses. Because of this, and because the metrical characters are relatively uniform, it is advisable to unite all the specimens under one designation.

That this species is not Scott's *Elephas zulu* is now clear. The latter bears considerable resemblance to the *Palaeoloxodon antiquus* of Europe, and is generally regarded as a variety of that species (cf. Haughton, 1932), whereas the specimens from Kaiso and Kikagati are more like the teeth from South Africa which Osborn described as *Archidiskodon subplanifrons* and *A. proplanifrons*, although they are narrower and of a more evolved type. They appear to be closely connected with some of the teeth described by Dart from the gravels of the Vaal River, unfortunately without regard to the known variability of such teeth and with a needless multiplicity of species.

Osborn (1934) published a partial revision of the South African specimens; reference to his table shows that the teeth from Uganda are to be placed in his "Group of *Metarchidiskodon griqua* (Haughton)", and it is better provisionally to identify them as that species, though with the generic appellation *Archidiskodon*, rather than to add to the super-abundance of names already proposed for the fossil elephants of Africa.

(ii) LOCALITIES OF POST-KAISO AGE

Kazinga Channel

Kobus kob subsp. This specimen (regd. M 15176) consists of the brain-case with a fine pair of horn-cores, together with the incomplete palate. It represents a fully adult male and was presented to the British Museum by Dr V. E. Fuchs. The right-hand horn-core is 380 mm. long, measured on the front of the curve, and 157 mm. in circumference at the base; the separation of the horn-cores at the tips is about 370 mm. These measurements agree well with those of an average head of *K. k. thomasi*, the Uganda kob.

Nsongezi

Two different collections[1] have been received from this locality. The first, presented by Mr Wayland, consists in the main of broken bones which were gathered from arbitrary levels at intervals of 1 ft. The second, collected by Mr O'Brien, is marked "N6–Nsongezi".

Mr Wayland's collection contains:

LAYER A (0–1 ft.). Carpel and tarsal bones of two large antelopes, one the size of an eland (regd. M 15177), the other the size of a harte-beeste (regd. M 15178).

LAYER B (1–2 ft.). Fragmentary bones of antelopes (regd. M 15179); one of them has been cut along the broken edges.

LAYER C (2–3 ft.). The distal end of a metapodial of a buffalo, and remains of small antelopes (regd. M 15180–M 15181).

LAYER D (3–4 ft.). Fragmentary bones of small antelopes (regd. M 15182).

LAYER E (5–6 ft.). Distal end of the radius of a young buffalo, and fragments of small antelopes (regd. M 15183).

---

[1] Both collections are from the Nsongezi rock-shelter, of Wilton-Neolithic A age. T. P. O'B.

Appendix B

LAYER F (6–7 ft.). Two bones of a small antelope and two teeth of a much larger species (regd. M15184). The teeth are heavily encrusted with mineral matter.

Microlithic Layer, Rock Exposure, Third Terrace[1]

In addition to some teeth of hartebeeste and other antelopes (regd. M15185) this layer contained a single tooth of a hippopotamus (regd. M15186) and an unworn second lower molar of *Hylochoerus* (regd. M15187).

Mr O'Brien's collection contains the right maxilla of a serval (regd. M15188) notable for the large size of the alveoli for the canines, part of the left mandibular ramus of a large jackal (regd. M15189), two anterior lower molars of a wart-hog (regd. M15190), and part of the right mandibular ramus of a young gazelle (regd. M15191).

Neither collection affords evidence of other than a late date for the deposits in which they were found.

Walasi Hill, Chui Cave[2]

That the deposits investigated are not very old is proved by the presence of the African elephant, which has not yet been recorded from deposits dated as Gamblian or earlier.

Thos sp. An upper and a lower tooth of a large jackal (regd. M15192).

Panthera pardus (Linné). A large leopard is represented by an astragalus and pieces of three metapodials (regd. M15193).

Leptailurus sp. A fragment of a left mandibular ramus with the carnassial tooth broken (regd. M15194) may represent a species of serval.

Hystrix sp. A porcupine is represented by two pieces of the skull containing the upper incisors and two cheek-teeth of one animal (regd. M15195) and by some isolated lower incisors and cheek-teeth (regd. M15196). They show that the animal was smaller than the species of *Hystrix* living in Africa to-day, and agree in this respect with remains obtained by Dr Leakey from Gamble's Cave (Hopwood, 1931, p. 274).

---

[1] The 30 ft. terrace at Nsongezi: from Wilton-Neolithic A horizon on the terrace.
[2] Containing a Wilton-Neolithic B industry. T. P. O'B.

Thryonomys sp. One upper incisor (regd. M 15197).

Hippotigris quagga subsp. Eight teeth (regd. M 15198), the two ends of a metatarsal (regd. M 15199) and two phalangeal bones (M 15200).

Diceros bicornis (Linné). Part of a lower molar (regd. M 15201) probably belongs to the black rhinoceros.

Phacochoerus africanus (Linné). Part of a third molar (regd. M 15202).

Bovidae. There are several species of antelopes represented by isolated teeth (regd. M 15203, M 15204).

Procavia sp. Remains of dassies are not common; there are two maxillae (regd. M 15205), both charred.

Loxodonta africana (Blum). Only two pieces referable to the African elephant were found, an anterior milk tooth with two plates and talons (regd. M 15206) which measures $23 \times 16 \cdot 5$ mm., and the rearmost plate and talon of a later tooth of the deciduous dentition (regd. M 15207).

In addition to the specimens already mentioned, there is a large number of bones of rodents. Many of these remains are charred.

LIST OF LITERATURE QUOTED

DIETRICH, W. O., 1937. Pleistozäne Suiden-reste aus Oldoway, Deutsch-Ostafrika. *Wiss. Ergebn. d. Oldoway-Exped.* N.F. Heft 4, pp. 92–104, pls. iv, v.

HAUGHTON, S. H., 1932. On some South African Proboscidea. *Trans. R. Soc. S. Africa,* XXI, pp. 1–18, pls. i–iv.

HOPWOOD, A. T., 1926. Mammalia *in* The Geology and Palaeontology of the Kaiso Bone-Beds. *Geol. Surv. Uganda, Occasional Paper,* No. 2.

—— 1931. Preliminary Report on the Fossil Mammalia. Appendix C *in* Leakey, *The Stone Age Cultures of Kenya Colony.*

—— 1935. Fossil Elephants and Man. *Proc. Geol. Assoc.* XLVI, pp. 46–60.

—— 1937. The former Distribution of Caballine and Zebrine Horses in Europe and Asia. *Proc. Zool. Soc. London,* 1936, Part 4, pp. 897–912, pls. i, ii.

OSBORN, H. F., 1934. Primitive *Archidiskodon* and *Palaeoloxodon* of South Africa. *Amer. Mus. Novit.* No. 741.

UGANDA

Miles

0   20   40   60

Roads ━━━   Railways ┼┼┼┼

30°   32°   34°

Kajo Kaji   Opari   Morongole
                                                                              Kakamari
Laropi   Nimule   KARAMOJA
    Dufile
                                        Kitgum   Magosi
WEST NILE   GULU   Acholibur
Arua
                                                                              Moroto
    Wadelai   Gulu
                    Victoria Nile
    Pakwach   Lira
Mahagi Port Panyamur Murchison   Atura   LANGO
              Falls   TESO
                    Butiaba   L. Salisbury
Kilo   Bukumi   Masindi Port   L. Kwenia   Soroti
                    BUNYORO   Nakitoma   L. Kyoga   Kumi   BUGWERE
    Kaiso Kitoba Masindi   Siroko
Kasenyi Kiziranfambi Hoima   BUSOGA   Mbale ELCON
                    Kitoma   Namasagali   Bubulu   Kitale
    Nyambirizi Kagade   MENGO Moulamuti   Busembatia Tororo
RUWENZORI MUBENDE   Iganga   Busia Malikisi
    Ibanda Fort Portal Mubende   Bombo   Jinja NORTH
TORO   Kyegegwa   Mityana   Kampala Bugungu Mjanji KAVIRONDO
Kisindi Katwe L.George Nabakazi L.Wamala   Kitala Buvuma Sigulu I. CENTRAL
    Katunguru   Katonga   ENTEBBE   KAVIRONDO
Lake Edward ANKOLE   MASAKA Bukakata Masaka   Sese Is   Rusinga I. Kisumu
                                        Kiungu   Kendu
        Mbarara Rakai   Homa
        Gayaza L.Nakavali Kijanebalola Sango Bay LAKE   SOUTH Kisii
    Ntungamo Nsongezi Katera Kiebbe   Karungu KAVIRONDO
    Kitaga Nyakanyasi   VICTORIA
Kabale   Kagera Port Kyaka   Bukoba
    L.Bunyoni
        Gabiro
        Gatsibu   Musoma
RUANDA
    Kigale

30°   32°   34°

INDEX

Index

Index